2

AGRICULTURE ISSUES AND POLICIES

AGRICULTURAL INFORMATION TRANSFER SYSTEMS

Agriculture Issues and Policies

Additional books in this series can be found on Nova's website under the Series tab.

Additional E-books in this series can be found on Nova's website under the E-books tab.

AGRICULTURE ISSUES AND POLICIES

AGRICULTURAL INFORMATION TRANSFER SYSTEMS

DEVA B. ESWARA REDDY

Nova Science Publishers, Inc.
New York

For permission to use material from this book please contact us:
Telephone 631-231-7269; Fax 631-231-8175
Web Site: http://www.novapublishers.com

LIBRARY OF CONGRESS CATALOGING-IN-PUBLICATION DATA
Eswara Reddy, Deva B.
Agricultural information transfer systems / author: Deva B. Eswara Reddy.
p. cm.
Includes index.
ISBN 978-1-61761-857-4 (softcover)
1. Agriculture--Information services. 2. Agricultural information networks. 3. Communication in agriculture. I. Title.
S494.5.I47E89 2011
338.1'6--dc22
2010034023

Published by Nova Science Publishers, Inc. † New York

DEDICATION

I dedicate this book to my parents

Nagi Reddy
and
Ramulamma

CONTENTS

PREFACE

VISTAS IN AGRICULTURAL INFORMATION TRANSFER SYSTEMS

The decision making process in agriculture rests squarely on information available to farmers, ranchers, entrepreneurs and policy makers. Information can best be considered as a productive resource, potentially limiting and influencing the efficiency of production. The agricultural information transfer system consists of four independent, interrelated components: development, documentation, dissemination, and diffusion of information. They broadly correspond to generation, organization, communication, and utilization of information. These components facilitate interaction, networking, feedback and collaboration by serving each other in a dynamic dual function as both a resource base and customer base. Each of these components has several subsystems. The advent of digital sources of information with electronic accessibility has made a significant impact on the information transfer process. By and large, information professionals focus on organizational and retrieval aspects of agricultural information such as abstracting, indexing and bibliographic services backed up with document delivery. Information is crucial at every level to ensure efficient production and workable marketing systems. The logistics and structural frameworks vary greatly among nations. International organizations such as the Consultative Group on International Agricultural Research (CGIAR) with location specific projects in various countries serve as information catalysts for global benefit. This book presents a holistic view of different components of agricultural information transfer systems with special reference to electronic accessibility to information and globalization.

ACKNOWLEDGMENT

Author would like to thank Dr. Irwin Weintraub for discussions of agricultural information dissemination issues in developing countries.

Chapter 1

KNOWLEDGE AVAILABILITY AND AGRICULTURAL PRODUCTIVITY

INTRODUCTION

Food is the basic requirement in the hierarchical needs of humans, the other two being security and the means for improved standard of living. It is more than 10,000 years since humans, in their attempt to ensure more reliable food supplies, changed from food gatherers to food producers by domesticating plants and animals. 270, 000 plants are known to science of which 7 000 have ever been used for food of which120 are cultivated today. Of these 120 plants, 90 cultivated plant species provide 5% of human food, 21 species provide 20%of human food, and 9 species provide 75% of human food (Harrison et al.,2001). Fewer than 2 million species of the working estimate of 13.6 million have been cataloged. There are now only about 30 plant species whose individual world production exceeds 10 million tons per year and six animal species whose production in the form of meat exceeds one million tons per year. Such dependence of the growing world population upon a few species has increased the vulnerability of food production systems to the hazards of weather aberrations and pest epidemics (Swaminathan et al., 1979). Food is an individual family problem and it is not enough for a nation to simply produce more. Each family needs to be able to produce or earn access to its food supply. Despite spectacular advances in science and technology and apparent economic prosperity, over one billion people worldwide are desperately poor. More than 800 million people do not have enough food or access to other primary commodities to meet their basic needs. The World

Bank estimated that 75% of the very poor live and work in rural areas and depend on agriculture for their livelihood, either directly or indirectly. More than one billion people still live on less than US dollar a day. Conversely, less than 4% of the population in the industrial countries and less than 2% in the United States are directly engaged in agriculture. While industrial countries face the problem of over production, developing countries face the challenge of increasing food production and ensuring food security. Since 90% of the food consumed in many developing countries is produced locally, increased production could improve the health and well being of the poor. Food security is the foundation for social security. In September 2000, in recognition of the need to alleviate poverty and hunger in developing countries, 189 countries including the United States affirmed a set of international Millennium Development Goals-MDG (United Nations, 2000). The goals include eradication of extreme poverty and hunger; achievement of universal primary education; ensuring environmental sustainability; and developing a global partnership for development. In order to achieve the Millennium Development Goals the Task Force on Hunger made seven main recommendations with several facets under each of them (UN Millennium Project, 2005). The third recommendation of this report advocates increasing the agricultural productivity of food-insecure farmers. One of the facets reiterates the need for effective agricultural extension services. In other words, technology/ information transfer for advancing agricultural productivity.

Food Security

Global demand for food is increasing as the population continues to grow and increase in prosperity. Demand pressure is concentrated in developing countries, where demand for agricultural products is forecast to increase at an average rate of 2 percent per year from 1999 to 2030 (FAO, 2003). Food demand is also affected by a shift in diets, which is occurring in developing countries as a result of increased prosperity, urbanization and changing preferences. Populations in developing countries are tending to consume more livestock products, more fruit and vegetables, and fewer cereals than in the past. Meat consumption in developing countries is projected to increase by 44 percent per capita from 1997/99 to 2030 (Bruinsma, 2003). Aggregate agricultural output is being affected through the linkages of livestock production to the crop sector which supplies feedstuffs such as cereals and oilseeds (de Hann et al., 1998). Hunger is a physiological sensation associated with insufficient food intake, closely related to the concept 'food insufficiency', meaning that an individual or household sometimes or often goes without food. (American Dietetic Association, 1990). The most common definition of 'food security' is 'access by all people at all times to enough and appropriate food to provide the energy and nutrition needed to maintain an active and healthy life'. Widespread hunger and malnutrition persists today despite considerable growth in per capita food availability (Barrett, 2002).

FOOD CONSUMPTION

The developing countries are depending on plant-to-man food chain whereas the developed countrieson plant-animal-man food chain.

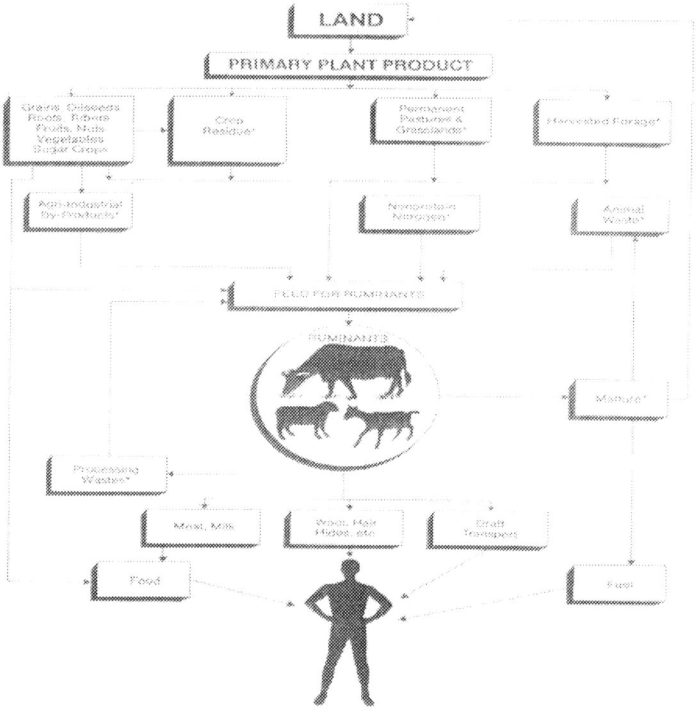

Figure 1.A graphic representation of land ruminant – man relationships. Products that are starred (*) are not consumed by man. Courtesy: Winrock international. From "*the Role of ruminants in support of man*" 1998.

It is conceivable that man can live without animals. It is inconceivable that animals and man can live without plants as depicted in Figure 1.

Consequently, the average per capita food consumption is about one ton of grain per year in the developed countries out of which only 70 kilo grams (Kg) are directly consumed by humans in the form of bread, biscuits and other products from flour. The remaining 929 kg are used to feed animals whose products like meat, milk and eggs are consumed by humans. In contrast, in developing countries the per capita consumption of grain per year is about 190 kg, most of which is directly consumed. Consequently agricultural productivity is crucial in providing food and feed. Food situation in developing countries is a perennial subject of international concern. Average food consumption is barely sufficient to meet the minimum needs daily requirement of 2,100 calories. Added to this the developed countries have resorted to use food grains for ethanol production which has direct bearing on the food security in developing countries.

AGRICULTURE AND NATIONAL DEVELOPMENT

Agriculture is the largest sector of economic activity in developing countries and provides not only food and raw materials but also employment to a very large proportion of the population. Increased agricultural incomes create market demands for industrial consumption goods thereby providing stimulus for market development. As commerce develops, the tertiary sector also expands. If a country had enough mineral resources, it can export the same, and in turn import food, raw material and even capital. But many developing economies are not so fortunately placed. About 75 to 80 percent crop production has continued to remain traditional in character. Conditions of static technology and cultural practices established over generations pervade the scene. Crop production is restricted to a few months in a year and is dependent on vagaries of weather. Purchased inputs are small. Seeds used are locally produced. Farmyard manures are used to the extent available. Farmers lack knowledge about soil and water consumption and the land management practices are inefficient. One of the principle causes of rural poverty is the low productivity of land and labor. Agriculture is a location-specific activity to produce food, fiber and feed. However, agricultural research is increasingly a global undertaking and the continued productivity of farmlands rests to a large extent on joint efforts to uphold and improve yields. Agricultural research depends on inputs from a broad range of disciplines. Isolated researchers are finding it increasingly difficult to confront the numerous challenges in increasing food production alone. For instance, agricultural technologies are sensitive to local climate, soil and other attributes. Unlike many medical, mechanical or veterinary innovations that are applicable from one place to another, local adaptation is required for crop technologies. Failure to raise agricultural productivity in the zones of turmoil and development could lead to recurrent humanitarian crises and result in setbacks to the globalization process. In the words of Jawaharlal Nehru, the first Prime Minister of India, "everything else can wait but not agriculture." In an ever-expanding world, agricultural development cannot wait. While in many countries of Asia and Africa 2.5 to 10 workdays are needed to produce 100 kg of grain, the labor time needed to produce the same quantity is 3 hours in parts of United States (Swaminathan and Ramanujan, 1980). The difference in productivity of farm labor is, therefore, about 1 to 800 and this gap also represents the relative earning and purchasing power of farm labor. National income has grown rapidly but there has been little change in living standards of vast masses of urban and rural poor.

ACCELERATED AGRICULTURAL DEVELOPMENT

There are two sides of the food problem. One is to increase food production and the other to create purchasing power among the hungry. The increase in crop and animal production on the hundreds and millions of family farms in less developed countries could be achieved only by casting the development net to involve small farmers. The educated and farm operators with large holdings are generally quick to adopt any new know-how which increase farm productivity. To enable the farmers to participate in the market economy the following four requirements must be met simultaneously.

1. Productive and profitable technology must be available
2. Farmers must be instructed to adopt the technologies
3. Necessary inputs such as seeds, fertilizers, pesticides, vaccines, feed for animals and credit must be available when and where the farmer need them
4. There must be markets for farm produce

If any of these requirements is lacking, farmers will not switchover to new productive practices. Efforts in the past in several countries often have centered on individual components such as supply of credit, or seed, or extension services, with disappointing production results. The fault lies in the design or execution of programs, including policy components. The third and fourth requirement can be met by creating a market for the output, but the first requirement has to be achieved by research and the second by putting the innovations to use or knowledge to work. Agricultural technology is largely biological, and therefore, must be developed for each season, of each region, of each country. Research will have to be done at regional stations and in farmers' fields to tailor the technology to existing farming systems. This demands decentralized field work by the nation's scientists. The individual farm is the ultimate experimental unit in any national system. The principal approaches for accelerating agricultural development are commodity programs, defined area campaigns, synchronization, and reorientation of services. There are two aspects to the problem of feeding the world's people. The first is the complex task of producing sufficient quantities of the desired foods to satisfy people's needs and the second task to distribute the food equitably. The chief impediment to equitable distribution is poverty – lack of purchasing power.

Agricultural Productivity

Most of the growth in population, labor productivity and real income per capita has occurred over the past 250 years. Improvements in agricultural technologies improve labor productivity and create surplus agricultural labor that can provide workers for the growing urban areas (Huffman and Orazem, 2001).Productivity growth in agricultural sector is essential if agricultural output is to grow at sufficiently rapid rate to meet the demands of food and raw materials that typically accompany urbanization and industrialization. Failure to achieve rapid growth can result in the drain of foreign exchange or in shifts in internal terms of trade against the industry (Hayami and Ruttan, 1970).

A substantial number of economic studies evaluating the contributions of research and extension programs in various countries to increased farm productivity and farm income have been undertaken in the recent past. A review of these studies by Evenson (2001) using estimated rates of return on investment to index economic impacts concludes that most of the estimates were consistent with actual economic growth experiences. Long history of public research in agriculture indicates that research benefits spill over to other regions and consumers, reducing the incentives for local support. A study of the productivity impacts of international crop genetic improvementresearch in developing countries over the period 1960 to 2000 concludes that international agricultural research centers, in collaborationwith national research programs, contributed to the developmentof "modern varieties" for many crops. These varieties have contributedto large increases in crop production (Evensonand Gollin , 2003).

POVERTY ALLEVIATION

Research-led technological change in agriculture generates sufficient productivity growth to give high rates of return in Africa and Asia and has a substantial impact on poverty, currently reducing this number by 27 million per annum, whereas productivity growth in industry and services has no impact. The per capita "cost" of poverty reduction by means of agricultural research expenditures in Africa is $144 and in Asia $180. By contrast, the per capita cost for the richer countries of Latin America is over $11,000 (Thirtle, Lin, and Piesse, 2003).

Improved cultivars produced by the international agricultural research system shows how new technology combines with other socioeconomic and institutional factors to determine poverty alleviation outcomes. Technology's role in alleviating poverty is both indirect and partial; technology alone cannot overcome poverty, nor can continued poverty be blamed on improved technology (Kerr and Kolavalli, 1999).

EDUCATION AND AGRICULTURAL PRODUCTIVITY

Schooling not only enhances agricultural productivity significantly following technology adoption but also promotes adoption it self as the study on the effect of farmer education on cowpea production under traditional and improved technology in northern Nigeria.

Factors that promote technology adoption will thus indirectly raise the marginal contributions of farmer education; these include schooling, participatory technology evaluation, improved seed supply, and market access. (Alene and Manyong, 2007).

The impact of primary schooling on crop production based on the first nationally representative household survey of Uganda was comparable to the developing country average (Appleton and Balihuta, 1996). Most of the studies on the subject have established that the education and skills of the agricultural workers are significant factors in explaining the inter-farm, inter-regional and inter-country differences in agricultural performance, along with the availabilities and potentials of natural resources of land and water, and infrastructure and institutional investments in inputs, credit, research, etc.(Singh,K. 2000).

Globally, different countries have evolved both unique and complementary systems to serve the science and education needs of their agricultural and rural communities. (Miller,Hanson, Fretz, and Weismiller,2004.)Modern agriculture depends on a continuous stream of new knowledge if it is to remain productive, competitive and profitable. Similarly, rural communities need to have well founded education and infrastructure systems to be viable and provide opportunities for their citizens to enjoy economic benefits. Schooling encourages innovation, not only directly but also indirectly, through its effect upon attitudes toward risk.

Educated farmers are early innovators and are copied by those with less schooling. (Tacchini,1995). A study covering a 30 year period 1965-1995 on the degree to which educational policies affect agricultural productivity in Kenya concluded that increases in education have had and will continue to have a large positive impact on agricultural productivity (Knight, Weir, and Woldehanna,2003).

Extension programs are positively affecting their service and recommendations to customers, both agronomically and environmentally, although not necessarily improving their business profitability. Extension will have greater impact on agriculture when it reaches key information disseminators directly (Schmitt, Durgan, and Iverson, 2000).

Information Transfer and Access

Information is a productive resource like land, labor and capital. At every stage of agriculture, the decision making process depends on formal and informal sources of information. Agriculture includes variety of operations, variety of resources. In a competitive world, information is the power. Sources of information consist of documents, institutions and individuals (Blackie and Dent, 1979).

The research process is not complete until adoption of new techniques by farmers, results in higher yields and increased incomes. Adoption requires that the farmer be made aware of new practice, become interested in it, try it out, and then adopt. Continuous flow of economically viable technology from laboratory to farmers' fields is essential for sustaining agricultural advances and for imparting stability in yield. The twin factors that have helped to raise the productivity in most industries have been the proper application of advances in science and technology and making adequate provision for the availability of thought resources of science and technology at the right time and in right manner to those who can make use of them for further advances in the field.

While the first factor needs careful planning, programming and execution by the planners and policy makers, the second factor of making available the thought resources is the concern of information professionals and extension personnel who have equal responsibility in the economic development of the nation. The second factor precedes the first factor, for the application of advances in science and technology is possible only when one is aware of such advances.

CONCLUSION

Agriculture is an interdisciplinary subject as well as interactivity subject. It is concerned with various resources such as land/soil, water, climate, such as seeds, fertilizers, pesticides/insecticides, capital, labor, equipment; operations such as tilling, sowing, mulching, harvesting; technologies for production of crops, livestock and fish, farm management, processing of crops and other agricultural products and disposal of agricultural wastes.

It is concerned with a diverse user community that involves such as government decision-makers, policy makers, planners, researchers, teachers and students, program managers, field workers, farmers, ranchers, business community as indicated in Figure 2.

All of them need different kinds of information for different purposes. An agricultural information system is a process in which information is generated, transformed, transferred, consolidated, received, and fed back in such a manner that these processes function synergistically to underpin knowledge utilization by agricultural producers (Röling, 1988).

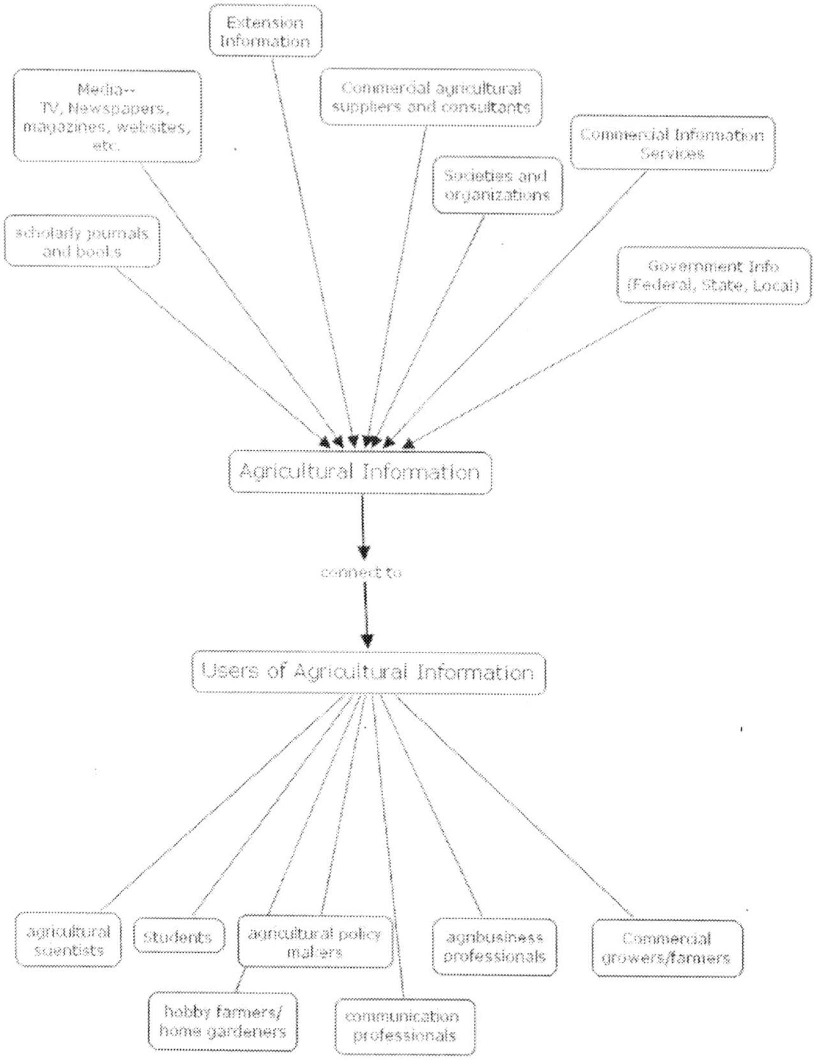

Figure 2. Concept map for agricultural information from Stocking the Shelves (Collections) – Task group Draft Report – Courtesy Janet McCue, Cornell University.

Information transfer is creation, dissemination, organization, diffusion, and use of information (Achleitner, 1995). In package oriented library or information publishing systems, delivery of information package is the goal, whereas the information transfer system is concerned not only with the delivery of ideas but also with the impact of these ideas on users (Landau et al., 1982). Briefly,the agricultural information transfer system consists of four independent, interrelated components: development, documentation, dissemination, and diffusion of information, which broadly correspond to generation, organization, communication, and utilization of information. The following chapter discusses these four components and their interconnectivity schematically.

Chapter 2

4 Ds Model of Agricultural Information Transfer System

Abstract

Information can best be conceived as a productive resource, potentially limiting and influencing the efficiency of production. An Agricultural information transfer system consists of four independent, interrelated components: development, documentation, dissemination, and diffusion of information. They broadly correspond to generation, organization, communication, and utilization of information.

Each of these components has several subsystems. Local, national and international educational and research systems worldwide are the prime movers in advancing agricultural knowledge. Organization and retrieval of agricultural information appears to be solely resting with library and information professionals.

Of late, these functions have also expanded to a plethora of national and international agricultural information systems. Starting with Extension or farmers' advisory service, the dissemination function will increasingly be oriented toward e-extension service worldwide. The fourth component of the information transfer system, utilization of agricultural information, has to overcome numerous barriers in meeting the information needs of farmers in developed and developing countries.

Introduction

The decision making process in agriculture rests squarely on information available to farmers, ranchers, entrepreneurs and policy makers. Information can best be considered as a productive resource, potentially limiting and influencing the efficiency of production.

Traditional communication systems reflect a one-way flow of information and the feedback is subtle. Information retrieval systems recognize the need for a two-way flow, with feedback, while in scientific information transfer there are multiple participants, with cross-communication among them. The National Science Foundation (1977) describes scientific and technical information as a spiral containing a sequence of functions performed by groups of participants--scientists/engineers, publishers, libraries, information centers, indexing and abstracting services--associated with documentation, dissemination, bibliographic control, storage and access. This model did not touch upon the feedback aspect.

The tripartite mission of teaching, research, and extension at US land grant colleges of agriculture (LGCAs) recognizes the vital role of information. However, the LGCAs model had no explicit place for information services in its mandate. Achleitner (1995) defined information transfer as creation, dissemination, organization, diffusion, and use of information. In package oriented library or information publishing systems, delivery of an information package is the goal, whereas the information transfer system is concerned with not only the delivery of ideas but also the impact of these ideas on users (Landau et al., 1982). The advent of digitized sources with electronic accessibility appears to have had a remarkable impact on the information transfer process. Models help describe how aspects of situations are related to one another. In other cases, models may assist in predicting how events will unfold (Britt, 1997). It is helpful to have a model to categorize the participants, their distinctive functions and their linkages with each other. This chapter presents a research model comprising the critical facets of interaction and feedback to facilitate an integrated approach to the components of agricultural information transfer. Figure 3, the Four Ds model shows how the agricultural information transfer system works. This system consists of four independent, interrelated components: development, documentation, dissemination, and diffusion of information, which broadly correspond to generation, organization, communication, and utilization of information. The participants in the Four Ds model facilitate interaction, networking, feedback and colla3boration by serving each other in a dynamic dual function as both a resource base and customer base. The 4Ds model serves as a conceptual framework within which linkages with four components are established. The solid lines illustrate the flow of information and the dotted lines show the feedback between and among the components.

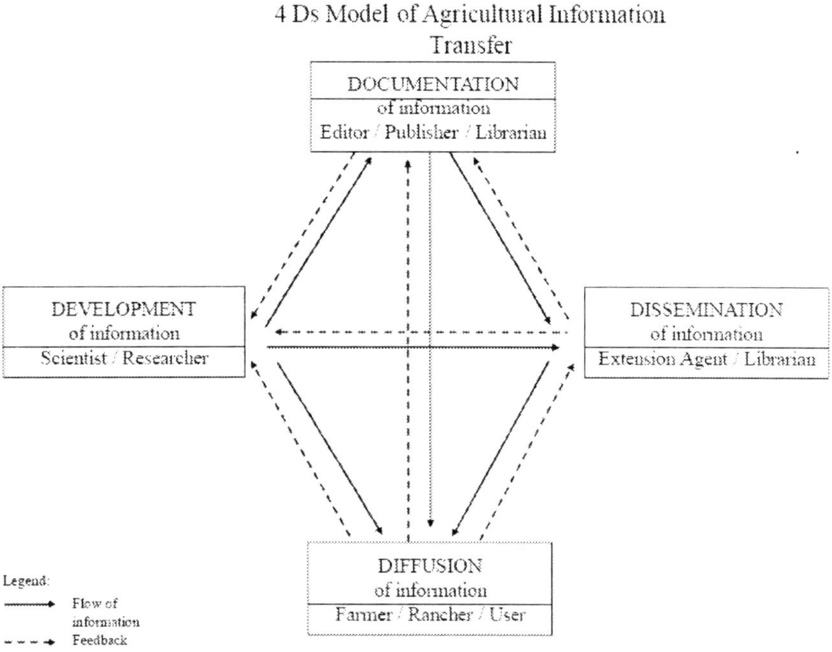

Figure 3. 4 Ds Model of Agricultural Information Transfer System.

For example, a progressive farmer may have direct contact with the researcher/scientist. The linkages between any of these components have both a functional and institutional meaning. They may be seen as functional links, which relate to research and information transfer activities; or as institutional links, which relate to the institutions and personnel that carryout these activities for exchanging resources such as, information,funding, labor, materials etc.

Development of Information

Development corresponds to advancement of knowledge and generation of information. Information is generated regarding new technology, improved plant varieties, soil types, crop production techniques, climate adaptability and recommended inputs to enhance sustainability. Such *information* is crucial for the farmers, the *agricultural* scientists and the overall development of *agriculture*(Sharma, 2003).A variety of institutions such as colleges, universities, experimental stations, national and international research institutes facilitate generation of information. The scientist as a part of his/her job sees, reads, writes, talks, and listens, which correspond broadly to observation, experimentation, analysis, interpretation, literature review, discussion and communication of results. Formal and informal channels of communication are extensively used to interact with fellow scientists, editors, publishers, librarians, change agents, and end users. Conventional and non-conventional sources of information including electronic sources are generally used at every stage of research. Also, the scientists are authors as well as end users.

Organization of Information

Documentation corresponds to organization of information. The editor/publisher helps the scientist with manuscript preparation, advises on publication ethics, and peer review, as well as interacting with the producers, distributors, and consumers of science information in identifying and correcting deficiencies in current publication processes. The editor may act as the author's mentor (Shashok, 2001). Librarians acquire, organize, and provide bibliographic and full text information to the scientist. They are the resource bases for all the users in the system. Over the years several national and international agricultural information systems have been developed. They include AGRIS of FAO (Food and Agricultural Organization of the United Nations) and Agricola of National Agricultural Library (NAL). In addition, several crop, animal or mission oriented subsystems such as soil information system, seeds information system, forage information system,water management information system have emerged. FAO and NAL pioneered in providing open access to their databases in agriculture which helped accelerate the open access movement for full text access to international agricultural information . In fact, the land grant colleges were among the early innovators when they phased out print dissemination in favor of digital dissemination to a larger audience.

Dissemination of Information

Dissemination corresponds to communication of information. Teacher, preacher, journalist, extension agent and librarian are partners in communication of information. There appears to be every need to distinguish technology transfer from information transfer. Technology transfer is show-how of know how of do how. The reference librarian/extension agent acts as information intermediary and disseminates the required information to the user in anticipation as well as on demand. The information is repackaged in such a way as to be understandable by the end user. The process is interactive and provides for multidirectional flow of information. E-extension connects knowledge consumers with knowledge providers - experts who know their subject and can interpret it for the recipient of the information who has to apply it to a specific need. E-extension is not like any other search engine or information-based website. It is a space where university content providers can gather and produce new educational and information resources on wide-ranging topics with speed and accuracy.(http://about.extension.org/).

Diffusion of Information

Rogers (1962) carried out pioneering work on the diffusion of innovations. Diffusion corresponds to utilization and assimilation of information. Utilization aims at putting knowledge to work. It also means evolving strategies for overcoming barriers to utilization of information, understanding information needs, and designing information delivery systems. The end user or farmer seeks information for observational purposes, problem solving and decision making. Depending on need, the farmer may use a variety of sources such as farm journals, product brochures, friends and neighbors, research institutes, manufacturers, government organizations, farm organizations, dailies, marketers and the Internet. Depending on the commodity in which the farmer is engaged, they require information on equipment, planters, harvesters, tractors, irrigation systems and other products and tools. Farmers are important agribusiness clients who purchase farm equipment feed, fodder, seed, pesticides, insecticides, chemicals, and other agricultural products.. For example, seed companies such as DuPont, Advanta, and Pioneer are simultaneously users, producers, processors, and vendors of information. In addition, farmers are in ongoing contact with public and private research institutions who can provide advice and assistance regarding operation of the farm business. Some farmers also conduct research on their own plots which can be shared with other farmers in the region. The end user interacts with the scientist, the librarian and the extension agent by providing feedback. While a lot still needs to be explored, and experimented, the trend is to explore the use of the latest participatory web-based tools, for example Web

2.0 tools such as Blogs, Wikis, MySpace, Facebook, Ning,LinkedIn, Twitter, and other social net working tools to improve collaboration, share experiences and enhance the flow of agricultural information to benefit all players in the agricultural transfer system shown in Figure 3.

The Web offers the opportunity to reach a larger number people at remote locations who can share and exchange information.. The Internet brought democracy to electronic information flow by narrowing class and economic distinctions and offering an opportunity for the formerly disenfranchised to be heard (Weintraub, 2000). It is a symbiotic relationship in which the contribution of each player helps make the 4Ds model a workable information system with practical application around the world.

CONCLUSION

Electronic access to information has resulted in productive use of precious time at every stage of research. Some of the significant impacts for scientists include: timely sharing of data and intermediate results for collaboration and discussion by eliciting instant feedback from colleagues and fellow scientists; global participation and narrowing the digital divide; instant interaction with publishers, editors, change agents and potential users; wider scope for publication and dissemination of findings; enhanced visibility of authors. The integration of scientific agricultural centers through connection to the Internet has facilitated direct links between scientists and thereby effective collaboration.

The exponential growth of electronic sources of information has also has greater impact on extension agents and librarians. Although networked electronic sources are accessible to end users, for a variety of reasons such as affordability, availability and acceptability, it is expected that greater use will be made of change agents and libraries. The Internet has altered the expectations of library users and changed the service perspectives of librarians and other information providers (Lankes, Collins, and Kasowitz, 2000). Increased expectations from the heterogeneous group of patrons involved demand different strategies for searching diverse databases, and this will have the positive impact of increased demand for technical skills to make use of technology. For example, the functionally literate farmer is able to interact directly with the scientist. In other cases the information has to be repackaged by the change agent/extension worker to be transferred either directly or through other channels. More sources will lead to more options for sources, resulting in higher patron expectations, and a need for more reliance on new technologies (Tenopir and Ennis, 1998).The advent of the Internet, reinforced with a variety of web-based information services, has meant that the boundaries between the author, publisher, librarian and user are shrinking. At the same time, the gap between the haves and have-nots is widening. Access to digital sources and, more particularly, collaborative efforts in building up digital sources of scientific information, are therefore a most welcome sign in closing the gap between those who create and disseminate information and those who use it.

Chapter 3

DEVELOPMENT OF AGRICULTURAL INFORMATION: NATIONAL AGRICULTURAL RESEARCH SYSTEMS

INTRODUCTION

Development corresponds to advancement of knowledge and generation of information. Science, more particularly basic science, is the foundation of agricultural research. Agriculture is the process by which humans produce food, fiber, fodder, and raise animals by making the best use of local land and water resources under prevailing climatic and technological conditions. Agricultural development and advancement in any country relies to a great extent on scientific and technological advances in agriculture so that the country is fully geared to meet the present and future challenges. One of the essential components of an agricultural information transfer system is a research system sensitive to the needs of ever-increasing populations. Agricultural research is probably the first and most widespread form of organized research in the world, and in which both the most developed and developing countries are engaged. While most forms of research activity, such as in the field of medicine, has worldwide application, agricultural research, by its very nature, has to be regional (Arnon, 1989). No research finding can be adopted without studying the results and application under varying ecological situations with which the farmers of the world are faced. Agricultural research covers a wide range of phenomena. Its results are used by different kinds of people – other researchers, farmers, extension workers, input manufacturers, and so forth (Lockeretz and Anderson, 1993).A viable and successful agricultural research system should have a sound organizational structure, research priorities and strategies to match developmental goals, linkages between the research and development institutions, and sound mechanisms to deliver research results to the clients. Research is vital if agriculture is to prosper. A nation's success in extending the agricultural revolution to all its regions depends in large part on its ability to produce a substantial number of people who: (1) understand national goals and objectives; (2) command basic farming skills; (3) are well grounded in the agricultural sciences and technology; and (4) are oriented to fast action (Wortman and Cummings, 1978). A variety of institutions such as colleges, universities, experiment stations, national and international research institutes facilitate generation of information. Mudahar et al. (1998) discusses four kinds of agricultural research. Basic research, which creates new scientific knowledge but without immediate commercial application; Strategic research, which provides knowledge and technologies to solve specific problems that have wider applicability; applied research, which develops new technologies and innovations by adopting basic and strategic research; and adoptive research, which involves selecting and evaluating technological innovations to assess their performance in a particular agricultural system. Despite spectacular advances in science and technology and apparent economic prosperity, over one billion people worldwide are desperately poor. Concurrent with population and economic growth, global demand for food is increasing rapidly; this is particularly critical in developing countries. Agriculture and food system development is vital to their survival and economic growth. Improvements in crop yield and farm management practices in these countries have always depended on the generous international flow of planting materials, implements, and ideas. This chapter presents an overview of national agricultural research systems in selected developed and developing countries.

BACKGROUND

Modern agricultural research systems date back to the 1700s. In developed countries, scientists in universities, government and the private sector contributed significantly to building viable agricultural research initiatives. Before the 1960s, relatively little was invested in agriculture research in developing countries, particularly for food crops. Prior to that, European countries and the United States were the main sources of international assistance aimed at expanding exports from developing countries. The

Netherlands supported export crop research in Indonesia and Surinam, Britain supported research networks in India, Canada, Australia, and New Zealand, while the United States contributed to building research programs in the Philippines, Liberia, Cuba and Puerto Rico.

France and Belgium aided commodity institutes in Francophone Africa and the Congo. All these countries generally focused on cash crops in order to generate income for the colonial population (Bernstein and Johnson, 1994). The Food and Agriculture Organization of the United Nations (FAO), through its technical assistance, has made significant contributions to the development of national research capabilities in agriculture.

Over the past several decades, three models for organizing professional resources to solve agricultural and rural development problems have emerged. The first is the counterpart model, in which an individual scientist or expert is employed by a technical assistance agency to function in an advisory role. The second is the university contract model, in which a university in a developing country and a university in a developed country establish a special relationship. The third is the international institute model, in which international institutes are established with a global agenda focused on a specific commodity. The latter model evolved from the historical success of imperial institutes focusing on commodities such as tea, rubber, sisal and sugarcane (Ruttan, 1989).

However, there appears to be no one optimum method of organizing the system as the country's agricultural conditions, history, economic characteristics, and socio-political traditions and present conditions play a key role in the optimul organizational structure. Trigo (1987) summarized five basic organizational types: The Ministry Model where the research is organized in one or more line departments within the bureaucratic structure of the ministry; The Autonomous or semiautonomous Institute Model where the research responsibilities are within an administratively independent organization; The University Model in which research is closely integrated with education within the university context in which extension activities may or may not be part of the same structure; The Agricultural Research Council Model representing a variant of the autonomous research organization, emphasizing the coordination function rather than the direct implementation of research activities; Private Sector Research Organizations Model involving and researching technological inputs such as (1) seeds, agrochemicals, fertilizers, farm machinery, and veterinary products; and (2) crop-specific research associated with agricultural producer associations. Rivera (2006) reviewed four major frameworks that have been advanced in the past 50 years regarding research and extension, and education by international organizations such as the CGIAR (Consultative Group on International Agricultural Research), the World Bank, and the FAO (Food and Agricultural Organization of the United Nations) to promote agricultural development. These four frameworks are known by their acronyms:

- NARI – National Agricultural Research Institutes
- NARS – National Agricultural Research Systems
- AKIS/RD – Agricultural Knowledge and Information Systems for Rural Development
- AIS – Agricultural Innovation System

These models are said to be "ideal types" which are seldom found in isolation and may coexist at the national level.

UNITED STATES –THE 'TRINITY' OF TEACHING, RESEARCH AND EXTENSION SYSTEMS

An agricultural research system cannot be thought of in isolation. Education, research and extension are interwoven while each of these functions has its own purpose. They are often better known jointly as agricultural educatio, research, and extension systems. The teaching, research and extension system in the United States is the best exemplar of development, dissemination, and diffusion effort of agricultural innovations.

However, it is well suited for a country like the United States where the percentage of the labor force directly involved in agriculture has gradually dwindled from 41 percent in 1900 to 0.7 percent *(farming, forestry, and fishing)*in 2009, and a country where mechanization and large scale farming is the order of the day. The first agricultural experiment station was established in the state of Connecticut in 1816 at a time when agriculture was the leading industry both in terms of national productivity and proportion of population participating in it (Etzkowith, 2001).

A research system is devoted to finding solutions for food and agricultural related problems through various agencies such as U.S. Department of Agriculture (USDA) – including over 100 USDA laboratory locations; 74 land grant colleges and universities; other institutions of higher education, and several private research facilities. Land grant colleges and other universities are vested with the primary responsibility for accomplishing the education component of the system in providing formal agricultural education for future farmers as well as for personnel to work as researchers and county agents. The Morrill Acts of 1862 and 1890 established the land grant colleges system, and the Hatch Act of 1887 started a system of state agricultural experiment stations under the direction of the land grant colleges.

Initially agricultural research concentrated on increasing agricultural productivity – crops and animals- by mechanical innovations for planting and harvesting. Today, agricultural research is very much diversified and includes plant and animal protection from pests and diseases, and other hazards, decreased use of chemicals, creating sustainable agricultural systems, food safety and nutrition, and improving the standard of living of farmers. With the advent of globalization, agricultural research in the United States is spreading its purview worldwide mostly through private firms.

The United States has played a leading role in establishing agricultural universities in several developing countries. Congress created the Cooperative State Research, Education, and Extension Service (CSREES) by combining the USDA's Cooperative State Research Service (CSRS) and Extension Service (ES) into a single agency. CSREES is one of four USDA agencies that make up its Research, Education, and Economics (REE) mission area. The other three agencies are: Agricultural Research Service (ARS); Economics Research Service (ERS); and National Agricultural Statistics Service (NASS). The CSREES' unique mission is to advance knowledge for agriculture, the environment, human health and well-being, and communities by supporting research, education, and extension programs in the Land-Grant University System and other partner organizations. CSREES doesn't perform actual research, education, and extension but rather helps fund it at the state and local level and provides program leadership in these areas. The USDA-REE agencies provide federal leadership in creating and disseminating knowledge spanning the biological, physical, and social sciences related to agricultural research, economic analysis, statistics, extension, and higher education.

The CSREES was replaced by the National Institute of Food and Agriculture (NIFA) in 2009 (http://www.csrees.usda.gov/).NIFA is a U.S. Federal government body whose creation was mandated in the Food, Conservation, and Energy Act of 2008. It is intended to consolidate all federally-funded agricultural research, and will be subordinate to the Department of Agriculture.

National Agricultural Research System in China

Food security has been one of the main goals of China's national policy. The priority of research programs had been in basic staple food grains. However, it has been gradually moving to non-staple crops, livestock and other agricultural products. The Chinese Academy of Sciences (CAS) is a comprehensive center for natural sciences in China. Its mandate is basic research, part of which deals with agriculture. CAS conducts research in basic and technological sciences; undertakes nationwide integrated surveys on natural resources and ecological environment; provides the country with scientific data and advice for governmental decision-making, and undertakes government-assigned projects with regard to key S&T problems in the process of social and economic development; initiates personnel training; and promotes China's high-tech enterprises by its active involvement in these areas. The organizational structure of agricultural research in China can be considered as ministry model subdivided into extension and research. China's agricultural research system is a multi-ministry model. The first agricultural research institute in China, the Agricultural Experimental Station of Beijing was founded in 1906. Agricultural research in China centers around the research institutes of the Chinese Academy of Agricultural Sciences (CAAS), provincial and prefectural academies, and to a small extent an agricultural university research system. Agricultural research is mainly the province of a special system of research organizations, though important work is also carried out in the university and teaching institutions as well. The institutes of the Chinese Academy of Sciences, the Ministry of Forestry, and the Ministry of Water Resources and Electric Power also undertake some research relevant to agriculture. There are four sub-systems with different responsibilities. The Ministry of Science and Technology (MOST), formerly the state science and technology commission, is the comprehensive organization devoted to the coordination and management of civilian science.

The ChineseAcademy of Agricultural Sciences (CAAS)

The CAAS was established in 1957. Different from provincial agricultural research institutes, CAAS is China's national agricultural research organization, directly affiliated to the Ministry of Agriculture. CAAS has a strategic task of serving nation-wide agricultural and rural development and empowering farmers with science and technology. Its major mandates focus on strategic and applied research solving key scientific problems which are of national or regional importance. CAAS is the only agricultural research academy in China that can confer a Ph.D. degree in agricultural sciences. CAAS has a Graduate School, five state key crop variety improvement centers and sub centers, 22 national and ministerial level key open laboratories, and 17 national and ministerial commodity quality supervision and testing centers. CAAS owns about 100 scientific and technology based enterprises with nearly 2100 personnel engaged in science and technology business development. CAAS is a key open window representing China's agricultural science and technology for international collaboration .CAAS has extensive international cooperation relationships with more than 40 countries and international organizations. Designated by the Ministry of Agriculture (MOA),CAAS plays a strong leadership role in China's collaboration network with the Consultative Group for International Agricultural Research(CGIAR) in teamed efforts with agricultural universities and other agricultural research institutions.

(http://www.caas.net.cn/engforcaas/) In addition, the Chinese Academy of Forestry (CAF) and the Chinese Academy of Fishery sciences (CAFS) promote research in their respective fields through various research institutes operating under their jurisdiction. Research at the Provincial Level is conducted through provincial agricultural academies while the prefectural agricultural research institutes administered by the prefectural governments perform research. China has 58 agricultural universities and colleges with a total of 800,000 graduates; 365 agricultural secondary schools with a total of 1.3 million graduates; 2600 plus agricultural TV and broadcasting schools at county level (Huang et al., 2003).

NATIONAL AGRICULTURAL RESEARCH SYSTEM IN LATIN AMERICA

In most Latin American countries, agriculture is a major source of income, employment, and export earnings which contribute to general economic growth, poverty reduction, and the sustainable utilization of natural resources (Echeverria, 1998). The region as a whole is a net food exporter but poor consumers suffer the negative impacts of food-price inflation on their incomes which affects health and nutrition very much. In addition, international value chains and supermarkets are transforming domestic food markets, thereby posing serious challenges to smallholders in their ability to remain competitive (World Bank, 2008). In the late 1950s, Argentina and Ecuador were the first in the region to introduce a model of public agricultural research founded on a single national agricultural research institute (INIAs in their Spanish acronym). This model was adopted by the majority of Latin American countries, mostly with technical and financial assistance from foreign agencies and foundations and donors such as the United States Agency for International Development, the Inter-American Development Bank, and the World Bank. In addition, producer associations, research foundations, private sector companies, and agricultural universities have also contributed to technology development in the region. The INIAs concentrate on shortand medium-term production problems.

Brazil has a two-tier system of federal- and state-based agencies. Among the government agencies, the Brazilian Agricultural Research Corporation (EMBRAPA) plays a vital role. It is a semiautonomous federal agency administered by the Ministry of Agriculture and Food Supply. EMBRAPA is the largest agricultural R&D agency in Latin America headquartered in Brasilia and operates 37 research centers throughout the country. The Research Center for Cacao (CEPEC) and the Brazilian Institute for the Environment and Renewable Natural Resources (IBAMA) are two additional federal agencies involved in agricultural R&D. In addition, 16 of Brazil's 26 states operate agricultural research agencies, although most state-level activities are carried out in São Paulo. The state's Agency for Agribusiness Technology (APTA) coordinates all São Paulo's crop and livestock research activities organized under four strategic programs: bioenergy, environmental sustainability, organization of the rural and outlaying areas, and food safety. Several universities in Brazil conduct research at more than 100 faculties or schools of agricultural sciences. Also a few of the private universities undertake agricultural research in Brazil, and the nonprofit sector plays only a modest role (Beintema and Pardey, 2001). In Mexico, agricultural R&D is overseen by the National Institute for Forestry, Agricultural, and Animal Husbandry Research (INIFAP). Headquartered in Mexico City, INIFAP operates eight regional research centers (CIRs) spread over the country, as well as five national disciplinary research centers (CENIDs). Both at the state as well as the country-level, other government agencies conduct agricultural research in Mexico. Several faculties or university units are involved in agricultural R&D. In Argentina, the National Institute of Agricultural Technology (INTA) is major government agricultural R&D agency. INTA integrates research and extension, making it unique among S&T institutions. In addition to INTA, a large number of agencies under the National Scientific and Technical Research Council (CONICET) under take scientific research covering both agricultural and nonagricultural disciplines. The University of Buenos Aires and other universities conduct agricultural research. Echeverria (1998) observes that research organizations and governments in the region are facing two critical challenges of obsolete institutional structures and lack of stable funding for agricultural and natural resource management research. Recent trends indicate the emergence of a pluralistic institutional structure, a growing role for the private sector, and new mechanisms for research funding, improved efficiency and effectiveness of public research organizations, and strengthening of global scientific links (Byerlee,1998).

The majority of the LAC's international research is carried out by the CGIAR via partnerships that date back to the 1960s. Of the current 15 CGIAR centers, three are headquartered in the LAC region: The International Center for Tropical Agriculture (CIAT) in Colombia, the International Maize and Wheat Improvement Center (CIMMYT) in Mexico, and the International Potato Center (CIP) in Peru. In addition, 10 other CGIAR centers have established more than 20 offices in the region. Several other international and regional organizations conduct agricultural research in LAC. For example, the Caribbean Agricultural Research and Development Institute (CARDI) headquartered at the University of the West Indies in Trinidad and Tobago, is the main agency involved in agricultural R&D in the English-speaking countries of the Caribbean. The Agronomic Center for Research and Education (CATIE) is an autonomous nonprofit institution that focuses on agricultural and rural development and natural resource management on behalf of its member states, which include all the countries of Central America, as well as Mexico, Bolivia, Colombia, the Dominican Republic, Paraguay, and Venezuela. CATIE combines research, training, education, and outreach, and—through its graduate school—its research programs focus on forestry and agro-forestry systems, strategic inputs for sustainable agriculture, the valuation of natural resources and environmental services, and rural development. New technologies resulting from R&D investments have enhanced the quantity and quality of agricultural outputs, while at the same time enhancing sustainability, reducing consumer food prices, providing rural producers with access to market opportunities, and improving gender-based allocations and accumulations of physical and human capital within households (IAASTD, 2008; Stads and Beintema, 2009).

National Agricultural Research System in Africa

The agricultural sector is the backbone of African economy. Although Africa has an abundant natural resources base of land, soils, water, forests and wildlife, it continues to be in an impoverished state, unable to attain its development potential (FAO,2005). The reasons for little or lack of development appears to be many and varied such as proliferation of wars and conflicts; unsustainable use of natural resources; droughts; lack of rural infrastructure and services. Approximately 80% of all Africans depend directly or indirectly on agriculture for their livelihoods. Agriculture provides 70% of Africa's full time employment, one third of total GDP, and 40% of total export earnings. Thus, Africa's overall economic performance is linked to the performance of its agricultural sector (http://www.fara-africa.org/about-us/background/).

It is said that the beginning of European scientific interest in Africa started with the invasion of Egypt by Napoleon (1788-1802) with an accompanying body of scientists who studied the riches of its ancient past (http://science.jrank.org/pages/48819/Africa.html). During the colonial rule, the British French, Belgian, Portuguese, and German governments established botanical gardens, model farms, and experimental farms both in Anglophone and Francophone countries in Africa. These gardens served as agricultural research facilities mostly for screening and propagation of tropical export crops such as tea, coffee, cocoa, cotton. Later in 1900s, the colonial governments started experiment stations which eventually served as the agricultural research infrastructure. By 1920 most of the African countries had at least one experiment station. In order to coordinate and execute the activities of these stations, the British established regional research stations in each country. However, the French experiment stations were consolidated as tropical research institutes and were managed from France (Pardey et al., 1995).

A majority of African countries inherited their agriculture research infrastructure along with political independence in 1950s and 1960s. Research, during the colonial period concentrated on export crops and very little attention was paid to subsistence farming (Beintema and Stads, 2004).While some countries inherited very specialized research institutes that did not necessitate much effort , smaller countries were unable to sustain the system once the staff and resources were withdrawn. As such, the smaller countries concentrated on strengthening research infrastructure and replacing expatriate staff with national researchers, both of which were uphill tasks for resource starved countries (Roseboom et al., 1998). During the 1970s and 1980s reforms concentrated on national agricultural research and development by amalgamating disparate research activities within a single national agency and improving management practices. The government is the major conductor of African agricultural research while the private for-profit sector accounts for only a small share of total research effort (Chema et al., 2003). In most countries, agricultural research is carried out by government institutions, universities, research foundations, and research facilities sponsored by agro-industries. The basic structure of larger institutions comprises an administrative directorate, laboratories, and research centers and commodity - specific experiment centers. The national agricultural research systems (NARSs), advanced research institutions (ARIs) and to some extent international agricultural research centers (IARCs) develop improved technologies in their respective countries and coverage areas in Africa. In order to coordinate these activities the Forum for Agricultural Research in Africa (FARA) was formed in 2002. It is an umbrella organization bringing together and forming coalitions of major stakeholders in agricultural research and development in Africa. It is currently hosted by the FAO Regional Office for Africa in Accra, Ghana. FARA is one of the regional forums linked to the Rome-based Global Forum of Agricultural Research (GFAR).

(http://fara.infosysplus.org/fara/fara_index.php/tpl/fara_ainfo/URL_NAME/fara). The agricultural education system in Africa consists of postsecondary education, universities, colleges, technical/vocational schools, and non-formal educational organizations. Michelson and Hartwich (2004) are of the opinion that in much of Francophone Africa, the colonial approaches did not change significantly with the coming of independence. However, postcolonial Anglophone Africa has witnessed several significant changes by linking their university research programs in agriculture to agricultural research and extension organizations, on the lines of the United States' model of land-grant universities and colleges (Lele and Coffman, 1995).Over the years considerable progress is seen replacing expatriates with national researchers followed by improvements in the quality of their training by offering postgraduate degrees. In 2000, there were over 12,000 researchers working in the public agricultural research agencies in Africa. In addition, these agencies employed 1.2 technicians and 3.7 other support staff per scientist on average bringing the total number of full time equivalent staff at African agricultural research agencies to 72,000 (Pardey, Roseboom, and Beintema, 1997). The majority of international research initiatives in Africa is carried out by the CGIAR. Of the current 15 centers under the CGIAR, four are headquartered in Sub-Saharan Africa- International Institute for Tropical Agriculture (IITA), Nigeria; International Livestock Research Institute (ILRI),Kenya and Ethiopia; Africa Rice Center (WARDA), Côte d'Ivoire, and World Agro forestry Centre, Kenya. Most centers headquartered outside Africa currently have offices in Africa, and in some cases they have considerable research facilities with ample internationally recruited staff. The studies by Maredia and Raitzer (2006) on CGIAR and NARS partner research estimated that over 11 million hectares are currently planted to CGIAR-derived varieties in the Sub-Sahara region. Apart from adoption of improved varieties or biocontrol, there were few cases where adoption of research outputs has progressed beyond tens of thousands of hectares. (Beintema and Stads, 2006). Many of the private-sector R&D activities in Africa focus solely on the provision of input technologies or technological services for agricultural production adopted from the developed world. Many private companies contract government and higher education agencies to perform research on their behalf.

National Agricultural Research System in Israel

The history of scientific research in Israel is an integral part of the story of the return of the Jewish people to its homeland. Agriculture in Israel is comprised of plant crops, afforestation and gardening, raising livestock, and livestock products. Agricultural research played an important role during the British Mandate period and during the first two decades of the state of Israel (Katz and Ben-David, 1975). Organized agricultural research began in the 1920s with the establishment of Jewish agricultural research stations, founded on the lines of the American land-grant colleges' model. In the 19th century, the farming communities were mainly the Arab fellah doingdryfarming, which was practiced for more than 2000 years, on a subsistence level (Loebenstein and Putievsky, 2007).A fellah is a peasent or agricultural laborer in Arab country, such a Syria or Egypt. Israel has intensively developed its agriculture and industrial sectors over the past 25 years. Out of the total labor force of 2.68 million, 2.6 % are employed in agriculture, forestry and fishing. Israel produces 93% of its own food requirements, supplemented by imports of grain, oil seeds, meat, coffee, cocoa and sugar.

(http://www.mfa.gov.il/MFA/Facts+About+Israel/Israel+in+Brief/).

The Agricultural Research Organization (ARO), of the Ministry of Agriculture and Rural Development, is responsible for most of the agricultural research conducted in Israel. This research aims to improve existing agricultural production systems and to introduce new products, processes and equipment, thereby ensuring the basis of Israel's future agriculture. The ARO has an extensive research infrastructure that supports both basic and applied research, conducted by more than 200 Ph.D. scientists and 300 engineers and technicians. (http://www.agri.gov.il/en/home/default.aspx). The Faculty of Agriculture, Food and Environmental Quality Sciences of the Hebrew University in Rehovot is a major partner in the conduct of agricultural research. Several regional research centers are operating, the major ones being the Northern R&D station, Southern R&D station and the Arava Valley R&D station. Israeli agriculture is better known for pioneering agricultural biotechnology, trickle-drip irrigation, soil solarization and the sustained use of industrial waste water for agriculture. Computers in agriculture are widely used such as computer-controlled irrigation systems, including the drip method, which directs water flow straight to the root zone of plants. Electromagnetic treatment of water to improve animal health and crop yields is impacting agricultural productivity. Through a well-established extension service system, research results are quickly transmitted to the field for trial and implementation, and problems are brought directly to the scientist for solutions.Israel is also a leader in the development of mechanized systems used to speed up harvesting and other operations.

National Agricultural Research System - United Kingdom

The International Agri-Technology Centre (IATC) describes Great Britain as a country whose widely diverse landscapes have sustained and nurtured strong traditions in agriculture. As a trading nation, Britain exports livestock, crops, machinery and equipment, technology, and research skills all over the world and imports what it cannot grow for itself. The United Kingdom of Great Britain and Northern Ireland covers some 24.5 million hectares. http://www.theiatc.org/page.asp?section=43&sectionTitle=UK+Agriculture

About 25% of Britain's land is arable, and almost half is suitable for meadows and pastures. Its agriculture is highly mechanized and extremely productive; about 2% of the labor force produces 60% percent of the country's food needs. Agricultural Science in Great Britain consists of roughly five periods (Russell, 1966). The first phase starting from the later part of the 16th century to the end of the 18th century consisted of solo researchers like Francis Bacon who worked on specific problems out of enthusiasm and initiative without any support. During the second period at the beginning of the 19th century, agricultural societies were established aiming at solving farmers' problems. These societies established laboratories for soil analysis and field experiments. The noteworthy development was the establishment of Rothamsted Experimental station , the oldest agricultural research facility in 1843. The third period witnessed the development of agricultural teaching at university level supported by government funding for teaching. During the fourth period started before the First World War several state-aided research institutes were established. The Agricultural Research Council was established in 1931 by Royal charter as an advisory body to Agricultural Departments. During the fifth period following World War II and ending in early 1970, the Agricultural Research Council was made responsible for food research and management of research institutes. Slater (1951) opined that the ARC has taken the view that research cannot be directed centrally. The Agricultural Research Council (ARC) formally changed its title in 1983 to Agricultural and Food Research Council (AFRC). Until the 1990s basic research in agricultural and related topics was grouped under the umbrella of the Agricultural Research Service (ARS). AFRC was succeeded by Biotechnology and Biological Sciences Research Council (BBSRC) in 1994.The Biotechnology and Biological Sciences Research Council (BBSRC) is the UK's public funding body for research and training in the 'non-medical life sciences', and one of the seven Research Councils (RCUK) sponsored through the UK Government's Officeof Scienceand Technology.

(http://ec.europa.eu/research/agriculture/scar/uk_en.htm).

BBSRC carries out its mission by funding internationally competitive research, providing training in the biosciences, fostering opportunities for knowledge transfer and innovation and promoting interaction with the public and other stakeholders on issues of scientific interest in universities, centers and institutes. The Institutes conduct long-term, mission-oriented research using specialist facilities. They have strong interactions with industry, Government departments and others(http://www.bbsrc.ac.uk/.)

NATIONAL AGRICULTURAL RESEARCH SYSTEM IN SYRIA

Syria is a middle-income country, with an economy based on agriculture, oil, industry, and tourism. It is said that agriculture has been developing in Syria for some thousands of years. Formal agricultural research (AR) began within the first experimental farms at DeirElhajar and Kharabo, close to Damascus, in the early 1940s. After independence in 1946, limited AR activities were carried out by some units of the Ministry of Agriculture and Agrarian Reform (MAAR), along with their other main duties of administration, service, policy formulation, etc.. The Supreme Council of Sciences is the official agency for defining and implementing national scientific research policy in Syria. The Directorate of Agricultural Scientific Research (DSSR) was established in 1964 by the Ministry of Agriculture and Agrarian Reforms (MAAR). In principle, DSSR has the over all responsibility for formulating national agricultural research policy. (ICARDA, 1999). The national agricultural system consists of scientific and technical institutions. They include:

- Six directorates under the governance of MAAR – Directorate of Agricultural Scientific Research (DASR); Directorate of Soils (DS); Directorate of Water Irrigation and Water Use (DIWU); Directorate of Cotton Bureau(DCB); Central Directorate of Citrus Bureau(CDCB); Central Directorate of Olive Bureau(CDOB).
- The six faculties of agricultural sciences affiliated to the Ministry of Higher Education.
- Other scientific institutions involved in agricultural research to a small extent include Atomic Energy Commission of Syria, General Organization of Remote Sensing
- Some other agricultural institutions of whose mandate AR activities represent a relatively small part (otherMAAR Directorates: Steppe, Mechanization, etc; state enterprises for tobacco and sugar)

There are three international/regional AR organizations in Syria:

The International Center for Agricultural Research in the Dry Areas (ICARDA), established in 1977 and supported by the Consultative Group on International Agricultural Research (CGIAR). The Center has a world mandate for the improvement of lentil, barley and fib bean; all dry-area developing countries for the improvement of on-farm water-use efficiency; and a regional mandate for the improvement of wheat, chickpea, and farming systems. The Arab Center for the Studies of Arid Zones and Dry Lands (ACSAD), was created in 1971 by the Arab League. The Center, based at Douma, Damascus, on the same campus as DASR, DS and DIWU, covers research in water resources, soil, crops, animal production, social studies and planning). The Arab Institute for Forestry and Grazing (IFG) was established in 1959 in Bouka, (Lattakiaregion). Its main responsibility is training Arab students for awarding diplomas: two years after high school. The country study (ICARDA,1999) on agricultural research in Syria says that land resources (research stations, farms) of the NARS institutions cover most of the agroecological zones and are considered satisfactory by most of the NARS institutions. However, scientific equipment, computers, vehicles, communication equipment as well as libraries and documentation services are considered rather inadequate.

National Agricultural Research System in Australia

With a population of more than 21 million, Australia is the only nation to govern an entire continent and has a federal system of government. Australia is a major agricultural producer and exporter. The agricultural products include cattle, wheat, milk, fruit and nuts, vegetables, wool, barley, poultry, lambs, and sugar cane. Australian research and development (R&D) originated when the settlers tried in earnest to adapt European crops and methods of farming to food production in an alien environment. The early farmers identified problems needing solutions and petitioned the government to invest in R&D. (Radcliffe and Clarke, 1996).The state or provincial governments are not subordinates to federal government. The six major states are self governing. In addition two commonwealth territories enjoy limited self-government. Income tax is the major source of revenue collected by the federal government and returned to the states. Australia's agricultural research system approximates the research institute model (Alston et al.,1999). The key research agency at the Commonwealth level is the Commonwealth Scientific and Industrial Research Organization (CSIRO). It was initiated in 1926 as Advisory Council of Science and Industry to respond to national challenges such as pests, weeds, and diseases that were affecting agricultural exports. In 1949 it was reconstituted as an independent statutory authority and performs a broad range of functions. The departments of agriculture or their equivalents of state governments carry out much of the agricultural research and development. R&D corporations were setup as semi-autonomous bodies as a partnership between government and farmers, with funding provided equally by each partner. There are 16 corporations representing major commodities as well as research on land and water management. Almost all Australian rural R&D is performed in the public sector within 10 federally funded universities, by state and territory governments, or by national agencies (Core, 2009). Another key provider of rural research services at the Commonwealth level is the Cooperative Research Centers (CRCs). The CRC program was established in 1990 primarily to encourage collaboration in research and development between private sector and public sector research bodies and foster research concentration. The Universities are also significant suppliers of research services to the rural sector. In part this has come from the larger number of regional universities. It also arises from significant Commonwealth funding to higher education research. Australia helps developing countries through the Australian Centre for International Agricultural Research (ACIAR). It was established in 1982 to assist and encourage Australia's agricultural scientists to use their skills for the benefit of developing countries, and at the same time work to resolve Australia's own agricultural problems.

National Agricultural Research System in India

Life in India occurs mainly in her 638,365 villages. The country's economic structure is dominated by agriculture. Agriculture provides a livelihood for approximately 67% of the work force. Varying agro-climatic zones and a variety of soil conditions have facilitated the cultivation of almost all the cash crops and food grains. The Indian Council of Agricultural Research (ICAR) is the apex body for overseeing agricultural research in India. The ICAR was established in 1929 as the Imperial Council of Agricultural Research. After India gained independence in1947, the Council's name was changed to its present name, the Indian Council of Agricultural Research (Randhawa, 1979). Over the years, ICAR has played a significant role in the support of agricultural education, research, and extension in India. ICAR's mandate is to co-ordinate agricultural research and development programs and to develop linkages with related organizations at the national and international level. ICAR is also entrusted with the responsibility of fostering and supporting the growth and development of agricultural universities. Among the major scientific organizations of the country (i.e., the Council of Scientific and Industrial Research, the Indian Council of Medical Research, and the All India Council of Technical Education), ICAR is unique in having concurrent responsibility for both agricultural research and education in the country. A well-organized national grid of cooperative research has been operating in the country for over half a century (Mruthyunjaya and Ranjitha, 1998; Reddy,2005). In this system, four sets of institutions are involved: forty-six ICAR Institutes, ten Project Directorates, ninety All Indian Coordinated Research Projects, and forty-six agricultural universities are conducting agricultural research. ICAR Institutes and Project Directorates are headed by Directors and are directly administered by ICAR. The All India Coordinated Projects operate through ICAR Institutes as well as agricultural universities.

The Coordinated Projects provide a link between the Central and the State level Institutions, facilitating the coordination of research efforts and the maximization of research output through the elimination of duplicative effort. Because of this coordination, it has been possible to speed the progress of research and to focus attention on solving location-specific problems which have been acting as production constraints. The Institutes/Directorates/Bureaus have a mandate for working on single or selected groups of crops, animals, or commodities. Some of the Institutes have regional stations to cover diverse agro-ecological areas for developing area-specific technologies. ICAR has established twenty-seven National Research Centers (NRCs). These Centers use a team approach to enable scientists, under one leader, to concentrate research on one type of crop or animal, e.g., soybean, banana, equines and cold water fisheries.

The creation of the All India Coordinated Research Project (AICRP), under the ICAR system, is a landmark in the history of agricultural research in India. The AICRP is a mechanism for building a nation-wide, co-operative, inter disciplinary research network linking ICAR Institutes with a focus on commodities and species of national importance. The AICRPs also strengthen the research base in each agricultural university. Also participating directly or indirectly in agricultural research activities are other agencies, including generaluniversities,specilized institutes, organizations,and various Ministries/Departments of the Central Government; the National Institute for Agricultural Extension Management; the National Institute of Plant Health Management; the National Grain Storage Institute; and the Central Food Technology Research Institute. In recent years, private and voluntary organizations have contributed to the research effort in agriculture.However, they concentrate on production and protection aspects of specific crops such as cotton, sorghum, and peanuts. The vast network of ICAR, which include Institutes, Bureaus, National Research Centers and Project Directorates, has approximately 30,000 personnel, out of which nearly 7,000 are engaged in active research and its management. In addition, forty-six state agricultural universities employ about 26,000 scientists for teaching, research, and extension education; of these, over 6,000 scientists are employed in the ICAR-supported coordinated projects.

CONCLUSION

Publicly funded agricultural research is going through a serious crisis almost everywhere, not only because of declining budgets but also because scientific projects have been carried out in arigid, mechanical, and linear fashion with limited opportunities for interaction with farmers and private enterprises. The private sector has become more prominent in the fields of biotechnology and information technology, which are influencing agriculture. The modalities for providing extension services to resource-poor farmers are also changing. Non-governmental organizations (NGOs) are striving to relate the real problems of farmers to the academic perspective of scientists. NGOs also supply extension and education services to resource-poor farmers (Henriquez and Franca, 2004).In both developed and developing countries, the forces of globalization, market liberalization, technological progress, and evolving ideas on the role of the public and the private sectors are exerting immense pressure forchange.

Chapter 4

INTERNATIONAL AGRICULTURAL RESEARCH AND AGRICULTURAL INFORMATION

INTRODUCTION

Public sector research in agriculture has many different approaches. It takes place at colleges and universities, research institutes, industrial laboratories and on the farm. In most developing countries, the National Agricultural Research Service (NARS) suffers from several drawbacks: Shortage of well-trained science and technical staff ; lack of government commitment to research, budget cuts, and low level of funding for operational expenses; low salary levels; poor incentives for research; lack of well established research and extension systems; and inadequate research priorities. India, Indonesia and Mexico are exceptions (Tribe, 1994). In general, NARS have extensive but under-funded staff, diversified field facilities, and a profound understating of local constraints. An international communication network exists for those with the tools to access it, but many research and development systems lack even a single operating computer. Some are unable to fund current scientific journals. Most have massive amounts of essential scientific equipment that is inoperative due to a lack of parts and maintenance. Many countries have vast numbers of underemployed extension staff. Many lack sufficient funds to buy fuel so that scientists can visit off-station experimental plots and many developing countries have bureaucratic systems and procedures that prevent scientists from direct contact with farmers (Gamble, 1997). These obstacles to open communication and effective information exchange can hinder the progression of functional agricultural research among countries. Concurrent with advances in information communication technology and reinforced by the globalization and liberalization of trade and economic activities, informal organized collaboration is becoming increasingly common in international agricultural research. Networks have been established to test crop germplasm over a broad range of environments in order to explore ways of boosting the efficiency of fertilizer use, upgrade disease resistance in livestock, and identify socioeconomic obstacles to improved agricultural output. Continued productivity of farmlands on every continent rests to a large extent on joint efforts to uphold and improve yields (Plucknett and Smith, 1984).

INTERNATIONAL AGRICULTURAL RESEARCH CENTERS (IARC)

International agricultural research has contributed much to human well being and played a major role in making the Green Revolution in Asia possible. The phrase "Green Revolution" was coined by the former Director of USAID, late William David, to describe the breakthrough in wheat and rice production in Asia that began during the mid-1960s (Borlaug, 2001). Development assistance, including agricultural research support, was viewed as an important means of confronting the threat of communism (Hardin, 1994). The USSR's policy to increase the import of grain and soybean to feed livestock and increase the consumption of meat, milk and eggs resulted in a world food crisis. In turn, the U.S. government placed an embargo on exports. Due to global concerns about impending food shortages, the FAO issued an urgent call for increased international agricultural research. To feed the growing population, increases in food production and international agricultural research were deemed to be the need of the hour. With favorable market prospects, farmers and traders in exporting countries saw no major problem. It was apparent that there was a pressing need for adoption and testing of technologies locally, in contrast to traditional agricultural extension systems. The concept of mission oriented research institutes that focused on a single commodity gained momentum. In fact, the commodity centered Collaborative Research Support Program (CRSP), developed and supported by USAID in partnership with developing countries and U.S. land grant colleges and universities, collaborated with relevant IARCs in this mission. Thus, internationally conceived and funded agricultural research began in the mid-1940s (Runge et al., 2003). International agricultural research is a small but critically important part of the global research system. While such research fosters the advancement of agricultural science generally, its true importance derives from such factors as the extent of international spillover, the efficiencies of working across national boundaries for many research projects, and the sharing of information and materials (Anderson, 1998). Networking, however, is more than communication. It is shared interests and interaction, building on information received and provided, and creating a relationship and a knowledge base to enhance international research. International agricultural research has played a vital role as an instrument to advance agricultural production in developing countries. During the imperial regimes, it was based upon commercial-commodity research, such as rubber and poppy. With the emergence of the CGIAR and its 15 international agricultural research institutes(Figure 4), the focus has shifted to food-based crops. Because of these efforts, spectacular results have been achieved in increasing world food production. The CGIAR has been a unique instrument of international cooperation. It has contributed much towards the Green Revolution and has made significant impacts in reducing poverty by increasing employment, raising incomes, and lowering food prices. From the early 1970s to 1994, farmers and researchers in the developing world, with the CGIAR's help, increased average yields in food crops (CGIAR, 1996).

Other significant international agricultural research efforts include Australian Center for International International Agricultural Researchand Japan International Center for Agricultural Sciences.

CHANGING TRENDS IN AGRICULTURAL RESEARCH

Publicly funded agricultural research is going through a serious crisis almost everywhere, not only because of declining budgets but also because scientific projects have been carried out in a rigid, mechanical, and linear fashion with limited opportunities for interaction with farmers and private enterprises. The private sector has become more prominent in the fields of biotechnology and information technology, which are influencing agriculture. The modalities for providing extension services to resource-poor farmers are also changing. Non-governmental organizations (NGOs) are striving to relate the real problems of farmers to the academic perspective of scientists. NGOs also supply extension and education services to resource-poor farmers (Henriquez and Franca, 2004). Pingali and Traxler (2002) point out that the focus of agricultural research and development has shifted from the public to the private multinational sector due to three interrelated factors: intellectual property rights, rapid advances in biotechnology, and global openings for marketing of inputs and outputs. These developments have created incentives for private agricultural research, specifically in crop improvement. However, according to Pingali and Traxler, private sector research has concentrated on maize, soybeans, cotton and vegetables, but the two most important food crops, rice and wheat, have received minimal attention from the private sector. They indicate that the poor and marginal farmers are unlikely to reap the benefits of this research because conventional technologies have not yet reached them and there is skepticism about new technologies reaching them in the short run.

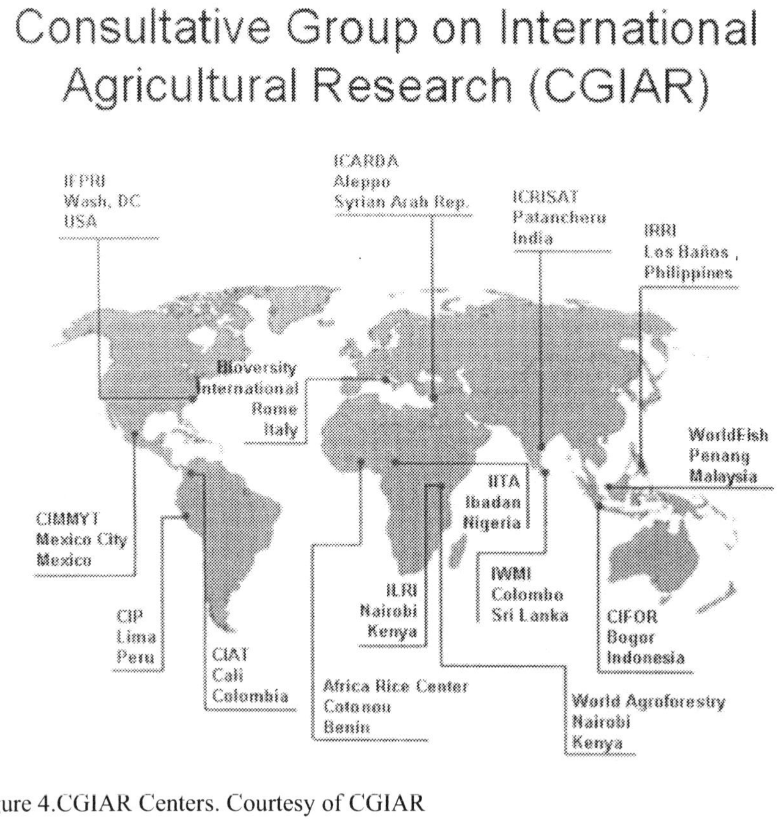

Figure 4.CGIAR Centers. Courtesy of CGIAR

In both developed and developing countries, the forces of globalization, market liberalization, technological progress, and evolving ideas on the role of the public and the private sectors are exerting immense pressure for change. Janssen and Braunschweig (2003) described the principal innovations observed in financing and organizing agricultural research in five developed countries-- USA, Australia, Switzerland, the Netherlands, and the United Kingdom; they also reviewed the policy environment to see how this has contributed to the reorganization of research. Knowledge generation and transfer mechanisms are being transformed in important and controversial ways. Investment in research and development has increased in response to the recognition that scientific productivity is tightly connected to economic dynamism. Patent protection has been expanded in order to stimulate higher levels of private investment. Wolf and Zilberman (2001) reviewed and analyzed contemporary transitions in agricultural knowledge generation and extension arrangements from an empirical perspective, and also emerging and contradictory perspectives as to how knowledge systems can be assessed effectively.

CGIAR AND THE CHANGINGCONTEXT

The CGIAR is facing huge challenges and is less focused on enhancing agricultural productivity than it was previously. Several factors explain the changing research mix and the increasing restrictions. First, germplasm improvement research has been unpopular among the constituencies of some key donors because of negative perceptions of the Green Revolution. Second, the CGIAR has responded to environmental regulations regarding soils and water created by the radical change in farming systems during the Green Revolution. Third, the growing environmental advocacy in donor countries has pressured the CGIAR to respond to environmental concerns. Fourth, the failure of many governments of developing countries and their donor supporters (including the World Bank) to make the necessary investments in their national agricultural research systems (NARS) has led CGIAR donors to turn to the CG Centers to fill the national and local public goods gaps closer to the farmer (CGIAR, 2003). The maintenance and management of CGIAR gene banks with 600,000 accessions of germplasm has suffered from inadequate funding. It has adopted a matching grant model instead of donor of last resort. In 2000, realizing that agricultural technology is only one instrument among an array of tools that can be used to attack poverty, such as infrastructure, health education, organizations, agrarian institutions and policies favorable to agriculture, CGIAR adopted a new "vision" of a food-secure world for all. It has decided to pursue regional approaches to collaboration with national and regional partners. By using regional approaches, both NARS and CGIAR derive mutual advantages by working together on behalf of national interests (Janssen and Kassam, 2004).

REASONS FOR CHANGE

Hardin (1994) identified some of the reasons why support for research increased rapidly, then declined. The factors that contributed support for the international agricultural research agenda include continuation of the cold war and national security concerns, and the possibility of food scarcity due to drought and famine in parts of Asia. The Soviet Union decided to expand the production of livestock, eggs, and milk by importing food grains to feed the animals, resulting in a doubling of prices. The U.S. placed an embargo on certain commodities. Due to population pressures and the growing urgency to meet food supply needs, improvement in local adoption of agricultural innovations was found to be crucial. CGIAR's rice research, and the U.S.'s CRSP worked well. However, it was followed by a decrease in research funding. Concern about the degradation of natural resources has moved to center stage with the disappearance of the cold war threat. Declining prices of basic foods and a rapid tilt towards privatization of agricultural research has also contributed to decreases in research funding. Finally, growing economic and political instability, civil discontent, and the terrorist attacks in New York and Washington further contributed to the growing unease over the relationship between the world's rich and poor (Runge et al., 2003). In the U.S., the food and agricultural sector is undergoing constant changes in the production, distribution, and consumption of food, fiber and technology to meet its own demands. These changes pose major challenges for public sector agricultural research and education. The Board on Agricultural and Natural Resources (BANR, 2002) identified the changes taking place in the U.S. agricultural sector, noting that those changes pose major problems in serving and meeting equally the needs of a large agricultural sector and multitudes of smaller producers, including low income and limited resource producers, and producers of niche commodities. The Board declared that publicly funded agricultural research has influenced the development of technologies biased towards changes in farm size and industrialization of the farm sector. In developing countries, there has been a drastic reduction in funding of agricultural research in the public sector, but there has not been a contrasting increase in the private sector.

CONTROVERSIES

Bill Christison (2000) is of the opinion that the increasing role of multinational corporations in controlling both the inputs and marketing of our commodities jeopardizes the existence of family-farm agriculture. There are two groups with quite contradictory opinions about development. The first believes that local self-reliance will ultimately prove the key to unlocking the shackles of the world's 800 million poor and hungry. This group is hostile to the global institutions, corporations and governments that dominate the international scene. The second group is in favor of private markets that forget about local conditions. Hunger and poverty are related to social justice at a number of levels. The global level seeks justice for developed and developing countries. National-level programs relate to distribution of income and humane treatment for a nation's poorer citizen. At the household level, which is designed to serve women, children, and the elderly, these groups strive to meet daily needs for food, clothing and shelter.

The concept of global public good is gaining importance in discussions on sustainable development. Public goods are often location specific. For example, flood control to alleviate the off-site effects of soil erosion and watershed protection can be referred to as a local public good. Public goods whose impacts are global in nature, e.g., climate change or migration, are referred to as global goods. When added to an already high demand in rich countries, rapid growth in demand for meat in developed countries has revived questions about whether using cereals to feed livestock contributes to shortages for humans. The answer is qualified. If humans did not eat meat, then some of the land area now devoted to feed grains could be redirected to food grains or other plant foods. Privatization and computerization of agricultural research leads to controversies about intellectual property rights (IPR) and the acquisition of patent rights. Patenting information results in gains at the cost of the taxpayer, while restricting its use hampers research and development as well as its adoption. In fact, this practice is now commonplace among several multinational corporations. So how can research and development (R&D) results be used for the public good? While some question the efficacy of public research organizations, others place greater trust in government-supported public sector research (Jain, 1999).

Implications for Information Transfer

The implications of the changing trends in agricultural research for information transfer appear to be manifold. A thorough understanding of these implications is possible when examined from a variety of approaches, e.g. the social, technological, environmental, economic, and political (STEEP) sectors. Shand (2001) asserted that giant transnational enterprises are gaining control over all aspects of commercial food, farming and health services. The dominant companies in plant breeding, pesticides, veterinary medicine and pharmaceuticals are known as "gene giants." Patenting genes, genetic traits and enabling technologies, legally restricts access to new agricultural biotechnologies. By the end of 1998, the U.S. Patent and Trademark Office had granted 1370 biotech patents to the top 30 patent assignees, 74% of which were held by six gene-giants: Monsanto, Dupont, Syngenta, Dow and Grupo Pulsar (Graff, 2002). Only 12% of corporate research focuses on farm level technologies. In contrast, 80% of public research is oriented to the farmer. By and large, corporate breeders are interested in industrial crop commodities such as soybeans, maize, cotton and canola, not in subsistence agriculture. On average, developing countries devote only 7.5% of total government spending to agriculture and less than one third goes toward research (Paarlberg, 2000). Annual foreign aid to agriculture in poor countries has been gradually reduced. Measured in U.S. dollars, the aid fell by 57% between 1988 and 1996, from $9.24 billion down to just $4.0 billion. (Alston, Pardey, and Rosebloom, 1998). With the decline of public sector research, who will address the needs of poor farmers, global food security, and environment, asks Shand (2001). Farmers who have been growing seed for generations now face legal barriers. It is said that up to 1.4 billion people in developing countries depend on farm-saved seeds as the primary seed source. Genetic engineers are building on the accumulated success of generations of farmers who have selected and improved seeds for thousands of years, whereas the private companies claim that they should have exclusive patent rights. For decades, public research institutions trained farmers and were the centers of agricultural research. About 45% of U.S. plant breeders at universities said that they have trouble getting seed from private companies, which places serious constraints on their research (Pollack, 2001). Information specialists are under constant pressure from two directions--changing agricultural policy as well as the time and costs involved in developing information infrastructures. The changing policies determine the types of information that researchers require and the ability of information units to respond to their demands (Ballantyne, 1994).

While CGIAR has a mandate for major food crops, several commodity crops not fully covered by the CGIAR mandate are very important to humankind. Some of these crops are essential for direct contribution to food security at the local and regional level, while others play a significant role in poverty alleviation and rural economies (Frison, 1999). Information services pertinent to such crops become crucial. Information service includes not only providing bibliographic information, but also service on demand by farmers. Who can do this? It has to be the specialists who reach out to the farmers. This is where the role of the extension service can have an impact, as there has been an apparent shift in research towards non CG-mandate crops

The transition from publicly dominated spheres of activity to one governed by markets, contractual arrangements, and property rights is referred to as a variant of privatization (Wolf, 1998). Wolf finds dramatic changes in the roles of public and private sector institutions engaged in development and dissemination of information in agriculture in the U.S. On one side, we have larger farms--highly capitalized, specialized farms, vertically integrated arrangements, and linkages between farms, input industries, and marketing institutions, in which the largest farms service their information needs internally, maintaining in-house expertise. On the other side are a growing number of large and medium sized farms without adequate information facilities. Due to the relative decline in public sector engagement in research, it appears that public extension specialists will play a less active role in direct delivery of information to farmers. Michael Boehlje (1998) looked at the problem from the technology and information transfer point of view. Privatization of information leads to less reliance on Extension and more dependence on private technical consultants. Obviously, this service will be out-ofthe reach for most small farmers in developing countries. Impartiality of fertilizer and pesticide recommendations made by private sector consultants could be at stake. Privatization tends to jeopardize innovative capacity to explore alternatives to dependence on fossil fuels and industrial inputs. Fewer farms with a concentration of assets will enjoy differential access to 'world-class-information,' leading to a geographical concentration of production in areas where the information infrastructure is well developed. Lack of public sector access to information internalized by individuals and firms will curtail the ability to conduct resource inventories, detect market failures, and undertake policy analysis. The positive impact of privatization of information presents opportunities for enhanced investment, efficiency gains, and accelerated innovations, especially in developing countries where public delivery of information is at a low level of access. Enhanced market development of analytic and diagnostic services, such as soil testing, integrated pest management, animal health, and technologies for fertilizer, pesticide, and water inputs will likely speed farm-level adoption. The implication is that agriculture scientists need broad knowledge of agricultural information in their specialty areas and related areas. Reliance on agricultural information resources that focus only on particular specialties may cause a knowledge gap that could hinder research efforts. Agricultural libraries in developing countries are isolated by virtue of their location and often limited to a small regional clientele. In addition, they often do not have the resources to build adequate subject collections. Most experiment stations are away from main campuses. In some countries one has to walk miles even to collect mail. Thus, major scientific breakthroughs that may considerably improve a nation's agriculture might not be apparent to extension agents without the use, dissemination and adoption of new technologies that disseminate information more widely.

CONCLUSION

In the Internet age, the costs of establishing and maintaining interactions with others, whether interpersonal or inter-organizational, are decreasing rapidly. The Green Revolution was greatly facilitated by the internal flow of information. Interactions between agricultural development workers and farmers were unencumbered by intellectual property rights and organizations like CGIAR that were vital parts of the information dissemination networks. However, the research and policy environment surrounding CGIAR has changed dramatically since its inception in 1971 due to revolutionary advances in biotechnology, the information and communication revolution, and the ascendancy of the private sector in agricultural R&D and the proliferation of IPR (Binenbaum et al.,2001). Failure to raise agricultural productivity in the zones of turmoil and development could lead to recurrent humanitarian crises and result in setbacks to the globalization process. Thomas Jefferson said that "knowledge is the common property of mankind." But in a market driven global economy, research results can be patented and the new products that research generates are private goods that produce revenue. If the clients are small farmers in the developing world, the very concept of global research may become archaic because many research tasks traditionally carried out by national and international research centers and program will be assumed by profit oriented enterprises.

Chapter 5

DOCUMENTATION AND OPEN ACCESS TO AGRICULTURAL INFORMATION

INTRODUCTION

In chapter two of this book, the third component of the Four Ds model 'documentation' has been conceptualized as the organization and retrieval of information. Webster's New College Dictionary defines documentation as the gathering, collection, synopsizing, and coding of printed material for future reference. However, information professionals largely refer documentation to indexing and abstracting services. We are besieged with ever-growing facts and information owing to rising standards of literacy, education and knowledge, proliferation of information producing organizations, and increasingly sophisticated means of communication and technology. In the field of agriculture, scientists and several associated individuals and institutions worldwide are generating information. They include research centers, experimental stations, academic institutions, commercial organizations, and marketing agencies. This information is accessed by the generators themselves, extension personnel, farmers and ranchers, and other consumers of information. This information is published in conventional and non-conventional form in a variety of publications. This includes grey literature too. With a view to having effective bibliographic control of the vast literature that is being published, national and international information systems have been established. However, the current trend is to provide for open access to scholarly information so that every one is able to access the information regardless of the countries they live in or professions they practice.

Open Access to Agricultural Literature

Open-access (OA) literature is digital, online, free of charge, and free of most copyright and licensing restrictions. OA was physically and economically impossible in the age of print even if the copyright holder wanted it. Internet and ICT have revolutionized the information transfer process. Open Access is public access. Public access has to be supported by public libraries, which in turn are run with public money. OA is free for users but costs producers. Some one has to pay for connection, infrastructure etc. The prestigious publications such as *Nature*, *Scientist*, and *Wall Street Journal* ranked Open Access among their science series along with genetically modified crops, human cloning, and global access to clean water. SPARC - Scholarly Publishing and Academic Resources Coalition- a coalition of academic and research libraries, has been a prominent advocate for open access. OA initiatives advocate free availability on the public internet, permitting any users to read, download, copy, distribute, print, search, or link to the full texts of these articles, crawl them for indexing, pass them as data to software, or use them for any other lawful purpose, without financial, legal, or technical barriers. Open access is consistent with the mission and goals of universities and societies to promote and disseminate information. Only small portion of research can be read without a subscription or paying for view an individual article. Most research is grant supported – taxpayers' money. Even the developed countries are not able to subscribe to all the journals. All are committed to advancing knowledge and technology transfer. A commitment to the value and quality of research carries with it a responsibility to extend circulation of such work to all who are interested in it and can profit from it. Globalization demands exchanging and sharing of information more particularly by multinational companies. In addition to peer reviewed journals, other vehicles for Open Access include self archiving repositories, OA journals, personal websites, eBooks, list serves, discussion forums, blogs, wikis, RSS feeds, peer-to-peer file sharing net works, and many more to come. Search engines are crawling open archive initiatives. Shift in OA will promote competitive edge between larger and smaller publishers. The origin of Open Access started with the French *Journal des scavans*in 1665. The National Education Library provided free access to ERIC database in 1966 followed by the National Medical Library for Medline database in 1997, Agricola of the US National Medical Library and, AGRIS of FAO of the United Nations. In fact, back in 1931, the eminent Indian Librarian, S.R. Ranganathan (1931) advocated open access in his classic book "Five Laws of Library Science". Most debate about OA is in science, technology, and medical sciences. Humanities and social sciences are lagging behind. The databases AGRIS, Agricola, and the CAB Abstracts cover the core areas of agriculture. There are quite a few other databases covering peripheral areas of agriculture. Most of the databases in vogue are accessible on subscription basis. However, with a view to providing access to everyone, only AGRIS and Agricola are open to public free of cost. In the field of health sciences, PubMed CentralBioMedCentral provide open access. For education and allied subjects, ERIC Database is freely accessible. Concurrent with the advances in information communication technology, several free websites, databases and institutional repositories have made inroads to the open access arena accessible via Internet.

AGRIS of FAO of the United Nations

With a view to facilitating information exchange and provide bibliographic control of agricultural literature emanating all over the world, AGRIS (The International Information system for Agricultural Sciences and Technology) was initiated in 1974 by the Food and Agricultural Organization of the United Nations (FAO). This database ((http://agris .fao.org) serves as comprehensive inventory of worldwide agricultural literature covering research results, food production, and rural development. The emphasis in AGRIS database is non-U.S. publications. The printed version of AGRIS is Agrindex, published monthly by the FAO of United Nations. Agris has adopted DCD model, in the sense decentralized input centralized processing and decentralized utilization. It is a cooperative effort in which over 135 participating countries are covered. In each country one institution is identified as a coordinating center which collects bibliographic information and abstracts of literature and submits to FAO for centralized processing the literature appearing in periodicals, books, reports, and conference papers. Approximately 130,000 records are added each year with keywords in English, French and Spanish. AGRIS database contains a great percentage of grey literature produced by governments, researchers, international organizations, research stations consultants and universities. In 2009, AGRIS has come under the umbrella of Coherence in Information for Agricultural Research for Development (CIARD), a global initiative of international partner organizations to increase the benefits from agricultural research and development.

AGRICOLA OF THE NATIONAL AGRICULTURAL LIBRARY

AGRICOLA (AGRICulturalOnLine Access) is a database created and maintained by the National Agricultural Library (NAL) of United States Department of Agriculture. The database serves as the catalog and index for the collections of the NAL and provides public access to information on agriculture and allied fields. AGRICOLA (http://agricola.nal.usda.gov/) indexes a wide variety of publications covering agriculture and its allied fields, including, "animal and veterinary sciences, entomology, plant sciences, forestry, aquaculture and fisheries, farming and farming systems, agricultural economics, extension and education, food and human nutrition, and earth and environmental sciences. The literature cited is primarily in English, but over one-third of the database comprises citations in Western European, Slavic, Asian, and African languages.

CURRENT RESEARCH INFORMATION SYSTEM

Initiated in 1966, Current Research Information System (CRIS)is the U.S. Department of Agriculture's documentation and reporting system for ongoing and recently completed research and education projects in agriculture, food and nutrition, and forestry. CRIS includes projects conducted or sponsored by USDA research agencies, state agricultural experiment stations, land-grant universities, other cooperating state institutions, and NIFA administered grant programs and allied projects. The major portion of the database is available free of charge to all users with access to the Internet (http://cris.csrees.usda.gov/).

CURRENT AGRICULTURAL RESEARCH INFORMATION SYSTEM

Created in 1975 by Food and Agricultural Organization of United Nations (FAO), CARIS is the Current Agricultural Research Information System to identify and to facilitate the exchange of information about current agricultural research projects being carried out in - or on behalf of - developing countries. CARIS identifies projects dealing with all aspects of agriculture. Over 30,000 currently active projects submitted by 137 national and 19 international and intergovernmental centresare described in CARIS. In addition, two regional centres submit input on behalf of 6 countries in their respective regions. CARIS is available free of charge to all users with access to Internet (http://www4.fao.org/caris/).

DIRECTORY OF OPEN ACCESS JOURNALS

The Directory of Open Access Journals (DOAJ) covers free, full text, scientific and scholarly journals and aims to cover all subjects and languages. There are now 5037 journals in the directory. Currently over 2072 journals are searchable at article level. As of May 2010, over 396434 articles are included in the DOAJ service. http://www.doaj.org/. A total of 279 journals are listed in this directory covering broadly agriculture and allied subjects.

GIVEAWAYPROJECTS

Realizing the importance of research and development literature in science, technology and medical disciplines for developing countries, several free for access projects have been launched. These projects are supported partly by the financial support of donors or patronage of publishers. For example, AGORA (Access to Global Online Research on Agriculture), HINARI (Health InterNetworkAccess to Research Initiative) administered by the World Health Organization (WHO), OARE (Online Access to Research in the Environment) coordinated by the United Nations Environment Program (UNEP) provide free access to certain countries with certain restrictions.

AGORA

The AGORA (Access to Global Online Research in Agriculture) program was launched in 2003 to help facilitate access to agricultural literature by selected developing countries with the support of FAO of United Nations and nine founding publishers. Other key partners include Microsoft and Cornell University. The eligibility for each country is based on gross national income (GNI) per capita provided by World Bank figures. Institutions in countries with GNI per capita below US$1250 are eligible for free access. Institutions in countries with GNI per capita between US$1251-US$3500 pay a fee of US$1000 per year per institution. Due to a publishers market interests and business plans, not all developing countries are eligible, as in some of these countries, the publishers have significant levels of existing subscriptions. Some of these countries include South Africa, India and China. As of March 2010, the initiative has increased to providing access to over 1278 journals.

AGRICULTURE NETWORK INFORMATION CENTER (AGNIC)

With a view providing free access for all those seeking agricultural information over the Internet, Agriculture Network Information Center (AgNIC) portal was launched. This portal is a voluntary alliance of over 60 member institutions and organizations. AgNIC (http://www.agnic.org/) is supported, in part, by the National Agricultural Library (NAL) of USDA. Based on quality and significance to AgNIC customers, AgNIC partners select important information sources that are readily available on the Internet for inclusion in the AgNIC system. Key information on selected topics covering a full range of agriculture, environmental, and food related disciplines is included. Members create subject sites such as rangelands, agricultural economics,acquaculture, subtropical foresrty etc. The AgNIC alliance is made up of professionals and subject experts from institutions and organizations from around the world. This is the best model for a cooperative effort of institutions and individuals in building up a dynamic knowledge base in a subject or group of subjects.

CGV Library

CGIAR supports 15 diverse international centers, located worldwide, generate scholarly research results that are considered as public goods. The CGIAR Virtual Library (http://vlibrary.cgiar.org)was launched in 2006. The CGVLibrary enables federated searching of catalogs and databases of CGIAR supported international agricultural research centers. This portal carries hyperlinks to full-text of CGIAR documents (Ramos et al., 2009).

Corporate Document Repository

As the description about the Corporate Document Repository (CDR) reads, CDR is FAO's online digital library. It stores thousands of documents in HTML and PDF format, containing valuable material on agriculture, food security and the other areas of work of the Organization. The CDR provides the users with two main search facilities, Free Text and Advanced Search, besides the general search engine located in the Home Page. Additionally, the CDR contains a specific section dedicated to the new documents which have been included in the repository in the previous month. The CDR is an open access facility and a great resource for agricultural research and development.

Coherence in Information for Agricultural Research for Development(CIARD)

With the advent of Internet and subsequent advances in information communication technology, a plethora of free websites, institutional repositories, and depositories are cropping up everyday. It becomes increasingly hard to keep track of these resources. Of course they are accessible and searchable via various search engines. Realizing the need for a voluntary registry of these resources, particularly in agriculture and allied disciplines, the Global Forum on Agricultural Research (GFAR) has initiated a portal 'Coherence in Information for Agricultural Research for Development (CIARD)'.The GFAR has developed a Routemap to Information Nodes and Gateways (RING) as summarised below based on the information from the website:http://ring.ciard.net/en/about-ring. The RING is designed mainly for agricultural information professionals and website developers, and secondarily for consumers of agricultural information as a "bookmark" list of agricultural information services. The RING is a web-based service that give access to information pertaining to agricultural research for development (ARD).It is the principal tool to allow information providers to register their services in various categories to facilitate the discovery of sources of agriculture-related information across the world. The RING provides a map of accessible information sources with instructions on how they can be searched effectively.

CONCLUSION

The rapid developments in information communication technology will bring new information services to agricultural scientists, extension agents, and farming community.These users will have greater control over current information channels. Every farmer may not have access to a computer but, could become readily available at local libraries and information centres in developing countries. What the farmer needs is the location-specific crop or animal or service oriented information service backed up with inputs such as seeds, fertilizers, pesticides etc. Concurrent with the developments in economic globalization, in the future, farmers will be better educated to be able establish direct contacts with the researchers and information providers.

Chapter 6

DISSEMINATION OF AGRICULTURAL INFORMATION

INTRODUCTION

Dissemination is a process by which something moves outwards from its point of origin including new ideas and innovations. Often the terms dissemination and diffusion are used synonymously, and refer to the spread of new knowledge or new techniques for those who need them. Dissemination can be called "planned information giving", whereas diffusion is "the spontaneous, unplanned spread of new ideas" (Rogers, 1983). Diffusion is not effective until the information is absorbed by the recipients. While industrial countries face the problem of over production, the developing countries face the challenge of increasing food production and ensuring food security. Since 90% of the food consumed in many developing countries is produced locally, increased production could improve the health and well being of the poor. The distinctive features of farmers are: They are not in the mainstream of academics; they are not located near academic institutions; they are not so functionally literate; most of them are not conversant with the intricacies of scientific literature. The farming community, unlike the industrial workers is not so well organized and unionized. Farmers need job related information; researched information; and basic information on crop and animal production.. Farmers, ranchers, and agricultural managers need to keep abreast of continuing advances in agricultural methods, both in the United States and other countries, as well as changes in governmental regulations that may impact methods or markets for particular crops. The agricultural productivity, profitability, and sustainability in the developing world depends on the ability of rural people in those countries to adapt to change and innovate in the use of technologies, management systems, organizational arrangements, institutions, and environmental resources. Innovation largely depends on their access to knowledge and information services. The Extension Service has proven time and again as the best means to provide advice to farmers on problems or opportunities in agricultural production, marketing, conservation, technology, and management of the farm business. Extension programs and services also provide rural inhabitants with issues of public interest such as resource conservation, health, nutrition, youth development and educational opportunities.

Communication of Agricultural Information

In the past, agricultural information was passed by word of mouth from farmers to farmers, friends and neighbors. With the advancements in science and technology, need arose to communicate information via different media. In today's environment, books, periodicals, libraries, agricultural organizations, newspapers, radio, television, government publications, house organs and a host of other pathways facilitate the flow of agricultural information from its source to those who need it. However, the development of the Internet changed the way agricultural information is disseminated. In the United States, the Agricultural Research Service of the US Department of Agriculture and the Cooperative Extension Service at land grant colleges became the primary providers of agricultural information by electronic means. Agricultural communications use mass media, electronic media, publishing, graphic design and information technology to extend the research of the governments and universities to public. However, the media mix is only for awareness. It will not solve the problem on land. According to the SMCR model (Berlo,1960; Boone et al., 2000), communication system consists of four important components - source, message, channel and receiver. These four components approximate to the four components in the Four Ds model mentioned in chapter 2 of this book. Source comprises the individuals or institutes generating information. Message is the content, which is the documented portion. Channel is the vehicle that disseminates the information, and the receiver is the farmer, scientist or anyone who uses that information. The communication products are the messages - they may be mass media messages, advertisements, news stories and radio or television broadcasts.

EXTENSION SERVICE

The term 'extension education' was first coined in 1873 by Cambridge University to describe particular educational innovations (Farquhar, 1962). Its mission was to carry the educational advances of the universities to ordinary people where they lived and worked. It is of interest to note that the first grants to the extension movement from public funds were for extension lectures in agricultural sciences. The American Extension movement started on the same lines as in England, from universities and large public libraries. Doctors (1969) states that the extension service has been established to operate in a rather unique environment which is characterized as a large industry with tremendous potential for technological advances in which no single member or group of members possessed either investment capital or the technological competence to develop or to utilize the needed technologies. The 1970 world conference on Agricultural Education and training treats formal education, vocational training and extension as three facets of information- exchange, transfer, and diffusion (FAO/UNESCO/ILO, 1970). The definition of a teacher includes teachers at all levels of agricultural education, technicians, trainers and extension workers. The fundamental objective of extension is to assist individuals in identifying and solving farm, home and community problems by utilizing the research findings. Brewer (2001) says that there has been no perfect extension system to meet every one's needs. Each system benefits from, and is constrained by, the environment within which it operates. Extension systems have been broadly categorized into five models.

1. Education Institutional Extension system: The U.S Land Grant System is an example of this model.
2. General Agricultural Extension systems: The extension service administered through the Ministry of Agriculture or Department of Agriculture is an example of an education institutional extension system. This is the most common practice in developing countries.
3. Commodity oriented extension systems: This system focuses on one crop, commodity or service such as coffee, sugar, tobacco, vegetables etc.
4. Non-governmental organization extension service: This extension system is designed around the specific objectives of nongovernmental organizations that work with but are independent of governmental agencies.

5. Parastatal extension systems: Owned or controlled wholly or partly by the government. The service provided by ADAS of United Kingdom is a said to be a parastatal service.

However, these systems overlap with each other. A worth noting extension system called Training and Visit system emerged in early 1970s and was implemented as the appropriate model in several countries in South Asia and Africa.

TRAINING AND VISIT SYSTEM

Realizing the need for improving transfer of farm technology, the World Bank expert Daniel Benor has designed an extensive extension service called Training and Visit system (Benor and Harrison, 1977). The World Bank in several developing countries financially supported this service in 1970s and 1980s. The thrust of the system is to have a systematic time-bound program of training and visits of field level extension workers combined with a clearly specified schedule of work and duties with regular and closer supervision. In this system, the transfer of knowledge from research station to the farmers in the fields is delivered in two stages. In stage one, the "Training," provides for transfer of knowledge from research scientist to extension workers through the subject matter specialists. In stage two, "Visits," provide for transfer of know-how from extension worker to the farmers. This model of Training and Visit system worked very well in India and several countries in Africa. However, with the curtailment of fundingby the World Bank, this model is appears to be no longer practiced in any significant setting.

AGRICULTURAL ADVISORY SERVICE IN THE UNITED KINGDOM

After the end of the Second World War, the British economy was stagnating. Food supplies and industrial production was very much affected. Before the war, Britain imported 55 million tons of food; a month after the war had started this figure had dropped to 12 million.

Rationing was introduced and the ration book became the key to survival for nearly every household in Britain. The critical need was to be more self sufficient in food supplies. It was recognized that increased food output could only be achieved by improving supply infrastructure centered around local shops, local supplies and markets, and minimum logistics. It was considered that such a major program could only be delivered by a state controlled development and extension system (Bunney, 1998). Technical advice to farmers began in 1890, following an Act of parliament providing for this service (Arnon, 1989). The National Agricultural Advisory service (NAAS) advising the Ministry of Agriculture Fisheries and Food (MAFF) was formed in 1946. Using MAFF Experimental Farms, Horticultural stations and commercial farms, the extension service was introduced by demonstration of technologies, farming systems, and techniques. The NAAS became the Agricultural Development and Advisory Service (ADAS) in 1991. In 1997, ADAS was privatized and is now known just as ADAS. To this day, ADAS remains a supplier to governments and a wide range of enterprises in the public and private sectors throughout the United Kingdom and internationally. ADAS operates from a network of 21 offices and research sites in England, Scotland and Wales with a head office in Wolverhampton. It employs over 400 people with expertise in all areas of agriculture. The Agricultural industry in the UK faces a variety of challenges such as increasing regulations, changing consumer demand, environmental protection, animal welfare, global warming, common agricultural reform and a worldwide concern with food availability and cost. ADAS provides farmers, growers, industry and regulators with expert help and advice on the most appropriate strategies to deal with these issues (http://www.adas.co.uk). The ADAS service, notionally commercialized operates on partial cost recovery basis. Clients of ADAS pay a fee for advice which formerly was free of charge. Reasons for privatization of agricultural extension include general discontent with the quality of governmental service, the decrease of financial support, and increase in dissemination of information.

AGRICULTURAL EXTENSION SYSTEM IN CHINA

Agricultural extension in China is said to have been in place over 3000 years ago. (Delman, 1988).The earliest beginnings of extension work in China was provided by a few private educational and missionary institutes and also by the then Nanjing University in 1920s (FAO/UNDP,1980). The beginning of the agricultural extension technology system (ATES), as it is now called, dates back to the1920s when sections resembling the Land Grant Colleges model of United States were established in many universities.

The objective of these stations was to carry out communications between the research institution and the farming community ((Qijie and Chuanhong, 2008). The ATE system is mainly led by the government and managed by the agricultural administrative department. Agricultural education, extension and scientific research belong to different government departments and they cooperate with each other. The ATE system is composed of five special extension organizations – agriculture, husbandry, aquaculture, agro-mechanization, and agro-business management. They are run separately. China's extension setup extends to all administrative levels from national, provincial, city, and county to township. The National Agro-Technical Extension and Service center formulates long-term development planning In rural extension, the county and town level extension organizations perform a major role. The tasks of extension system include experimentation, demonstration, training, extension, and liaison with farmers. In addition to public agricultural extension, other agencies such as R&D institutions, private companies, and producers' organizations provide extension services. Mission oriented or commodity oriented extension stations have been developed such as crop protection, livestock, tea, mulberry, and cotton stations. Agricultural extension is organized on the basis of these agricultural subsectors. With the abolition of the peoples' commune system, the Rural Household Responsibility System (RHCRS) was implemented in early 1980s. Under the RHCRS farmers have right to make independent decisions on using the small piece of contracted land. This has resulted in millions of clientele utilizing the services of extension instead of small numbers of communes, brigades, and production teams trying to handle it independently (Crook, 1992).

By the mid 1980s China had established extension stations in every rural county and township (Hu et al., 2009).The extension team consists of professional technological staff and farmer technicians. By the middle of the 1980s, the public agricultural extension system (PAES) employed an extension staff of more than one million at all levels; about 70% had graduated from technical high schools or colleges. More than 90% of them worked at PAES stations at the county levels. In some provinces, a system of contract extension is practiced under which a farmer or a group of farmers pay 20 percent of the value of the crop above the agreed target. If the harvest falls below the agreed target due to impractical advice or lack of assistance with needed inputs, the pay or bonuses of contracted extension workers is reduced to 80% of the shortfall. Research institutions, agricultural colleges, individual scientists, and teachers may sign contracts with rural units or farmers technical support on an annual basis or continual basis (Amuer, 1994). Farmers' associations render significant help in increasing agricultural production by expeditiously disseminating improved technology. However, they do it on ad hoc basis. It is said that contract extension in China is popular with the farmers because of the inputs they receive from the agents. In China each farmer is required to sell a specific amount of staple food such as wheat, rice or maize to the government at a fixed price. Since the average size of a farm is only equivalent to one-half acre of irrigated land/farm, this requirement is not easy to meet with. Depending on farm size and proximity to urban markets, some farmers concentrate on livestock enterprises (e.g., pigs, ducks etc), while others grow horticultural crops (vegetables and fruit crops). These growers then use some of the profits and buy the required amount of food staple crops / products from other peasant farmers (Swanson, 2006). About 95 percent extension workers work for households while the rest work with the large commodity oriented state farms.

AGRICULTURAL EXTENSION SYSTEM IN INDIA

In India, although agricultural extension service began with a single line extension service, with the introduction of community development programs in 1952, this was radically changed.

The community development program was followed by other agricultural development programs such as Intensive Agricultural District Program (IADP), Intensive Agricultural Area Program (lAAP), High yielding varieties program (HYVP), National Demonstration Program (NDP), Multiple Cropping Program (MCP), special programs for commercial crops - tobacco, sugarcane, cotton and jute, oil seeds - fruit crops and Dry land Agricultural Development program (DADP). Some of these programs are related to other allied sectors also.

A network of extension institutes has been created for the promotion and adoption of improved technologies among farmers. The research results are disseminated through the university extension officers to the departmental personnel as well as to the farmers. The ICAR has specific programs such as national demonstration program, operational research projects, *Krishivigyankendras* (Agricultural knowledge centers), lab to land program *kisanmelas* (Farmers fairs) etc. In most states in India the Director of Agriculture is the head of the department charged with the responsibility of formulating policies and programs and communicating them to the district level field staff for implementation and as such he is responsible for agricultural extension and development. The Director of Agriculture is assisted by one additional Director and several Joint Directors. At the district level the Deputy Director of Agriculture works with the assistance of two or three Assistant Directors of Agriculture (Subject Matter Specialists). Each revenue division is headed by one additional Director of Agriculture who in turn is assisted at the block level by Agricultural Extension Officers. The Agricultural Extension Officers are assisted by Village Development Officers and Field Assistants who convey the agricultural messages.

CURRENT TRENDS IN THE UNITED STATES

Farmers and ranchers are besieged by a wave of innovations associated with emerging technologies. Major shifts are occurring in the way information is accessed by farmers, including diversification of channels through which farm operators receive information. The United States federal Cooperative Extension Service (CES) began more than eighty five years ago as a way to disseminate agricultural research information from land-grant universities. Extension educators have recognized the value of distance delivery of educational programming, and they are exploring new ways to make extension a participatory learning organization (Kerka, 1998).

Extension education has traditionally focused on face-to-face, county-based programming where typical extension clientele have been rural residents who live on the land they own. Today, however, the increasing number of landowners who are not traditionally associated with Extension – specifically absentee landowners and part-time farmers – makes it difficult to reach them through workshops and meetings. In light of this fact, many states have begun searching for educational alternatives. The current dominance of the Internet, which is convenient and cost effective, has made it another logical delivery method for Extension programming (Jackson et al., 2004). While some farmers and ranchers may question the accuracy of information obtained from different sources, they trust the information from the CES because it is research-based and unbiased (Pates, 2004). Electronic media (such as email, the Internet etc.) is an excellent way to disseminate information from universities to county educators and producers (Latour and Meunier, 1999). According to the 18th annual survey conducted by the American Farm Bureau Federation (American Farm Bureau, 2010), the Internet is an important tool for young farmers and ranchers. Nearly 99 percent said they have access to and use the Internet, with the vast majority, 72 percent, saying they have access to a high-speed Internet connection. Only 20 percent rely on slower dial-up connections and 8 percent turn to more costly satellite connections. The social media site Facebook is very popular with young farmers and ranchers.

Nearly three-quarters of those surveyed have a Facebook page. Ten percent of the young farmers say they use the micro-blogging web site Twitter, while about 12 percent say they post YouTube videos. Communicating with consumers is also important, with 77 percent saying, they consider reaching out to the public about agriculture and their operations an important part of their jobs as farmers and ranchers. In addition, the Internet is an important tool for access to both general and farm news, with 84 percent saying they use the Web for that function. Seventy-two percent said they turn to the Internet to collect buying information for their operations. The Farm Computer Usage and Ownership report released in August 2009 by the USDA's National Agricultural Statistics Service (NASS) shows that a total of 59 percent of U.S. farms now have Internet access, compared with 57 percent in 2007. Sixty-four percent of farms had access to a computer in 2009, the same level as 2007 (National Agricultural Statistics Service, 2009).The World Wide Web (WWW) is being used extensively in agriculture to increase the availability of information to farmers and ranchers. Accessibility from anywhere and with minimum client-side capability is the key to allowing everyone to participate.

E-EXTENSION

E-Extension has been defined as the use of Web and Social Media tools to share and create agricultural and environmental knowledge (Thinke Extension) (http://thinkeextension.com/eextension/so-what-is-eextension/). eExtensioncan be defined simply as the use of electronic technologies (especially Information and Communication Technologies or ICT) to enhance face-to-face (f2f) and paper-based interactions. These technologies can be as simple as teleconferences or as complex as wikis and blogs.E-Extension in the United States, is an Internet-based educational partnership of the seventy-four 1862 and 1890 institutions of the land-grant university system.

The multiple objectives of E-Extension include providing expertise, reliable answers based upon sound research, establishing linkages with the best minds in American universities, creative solutions to today's complex challenges, getting answers to users' specific needs by providing reliable field-tested data, and dynamic, relevant and timely answers.

E-Extension is not a search engine like Google or Yahoo or information-based websites. It is a space where university content providers can gather and produce new educational and information resources on wide-ranging topics. It is available to students, researchers, clinicians, professors, as well as the general public, at any time from any Internet connection; extension helps solve real-life problems in real time. It is a 24/7/365 access to information to help people make life-improving decisions; educational products and programs at any time, from any place, in any format on any Internet-ready device. It was conceptualized in 1994 and went through various stages of discussion, deliberation, prototyping and, finally launched in 2007. Although it is deeply rooted in the United States the services are open to one and all.

Mobile Phones and Rural Communication

Although information communication technology (ICT) has revolutionized information transfer processes in every sphere of activity, the availability of a mobile phone has become a boon for rural folks, more particularly for farmers in remote areas in developing countries.

Rogers (1983) opined that the exchange of information and its diffusion take place within a social system. Actors such as individuals, informal groups, organizations and subsystems are the members of the system and the structure of the social system and their actors or members' roles affect the diffusion process. This concept was reiterated by the former president of the World Bank Robert McNamara (1977) who said that messages can be sent electronically thousands of miles, but it is ultimately people talking to one another in a classroom, on a street, at the village market, or in the village home where the essential questions are discussed and essential answers are explored. The mobile phone has greatly facilitated establishment of contacts for personal sources of information.

For example, subsistence farmers, would sell the produce not for profit but to get necessities like kerosene to fuel lamps in homes, sugar for making beverages, buying school supplies for kids, clothing, and other daily needs. They are now able to use the cell phone to obtain the latest market price. Many of these new mobile citizens as depicted in Figure 5, live in far-flung places with scarce infrastructure, and varying degrees of literacy. They face several constraints in using the handset. They would want rural-friendly mobile telephones to provide substantial battery-time, be dust and water resistant and improved signal sensitivity.

For greater ease of use, large laser buttons on keypad, adequate display features, and multi-lingual keyboards, voice messaging, alarm clock, and flash light, and FM/Radio should be offered for user convenience at a cost affordable by the rural folks worldwide. (Neelameghan and Chester,2010) Handset makers, such as, Nokia, Motorola, LG, Samsung, Sony Ericsson are examining and prioritizing the kind of facilities desired in mobile phones by rural folks, as differentiated from those by urban users (Krishnakumar, 2007).

Figure 5.A farmer in Indiausing a cell phone in the field on an ox.Courtesy – M.S. PrakashBabu.

Conclusion

The extension services have been criticized on several grounds such as being supply-driven, technically weak, patronizing only big farmers, insufficient coverage of and contacts with farmers, practicing top-down extension approach, etc. Some of the criticism is genuine while some has been levied without understanding the underlying causes which are beyond the control of extension workers, such as poor pre-service education, little in-service training, burden of non-extension tasks, low salaries, low status, lack of opportunities for professional career development in comparison with other agricultural services, and needed coverage of a very large number of farmers scattered over a very large area without operational budget or transportation facilities. Not surprisingly, the extension services have usually become the first victim of any major economic reform. For example, the number of public extension workers is drastically reduced due to structural adjustment measures, recommended by major donors. In addition, during the process of decentralization, extension services are marginalized and downgraded. Also, the remote areas, which extension agents rarely visit due to lack of transportation facilities, are increasingly being covered by NGOs. The global criticism has called for alternatives such as privatization of extension services, inclusion of other partners in extension delivery, contracting-out of extension work, and farmer-to-farmer extension modality. However, the extension model will continue to predominate in country specific settings as the need to grow food and fiber crops for expanding populations becomes a necessity.

Chapter 7

Holistic View of Agricultural Information Transfer System

Introduction

Diffusion corresponds to utilization and assimilation of information. Utilization aims at putting knowledge to work. It also means evolving strategies for overcoming barriers to utilization of information, understanding information needs, and designing information delivery systems.

The end user interacts with the scientist, the information professional and the extension agent by providing feedback. While a lot still needs to be explored and experimented, current trends make it possible to explore the use of the latest participatory web-based tools, for example Web 2.0 tools such as blogs, wikis, tags, and other social net working tools to improve collaboration and share experiences for the benefit of agricultural development. Although agriculture is a location-specific activity, agricultural research is increasingly a global undertaking and the continued productivity of farmlands rests to a large extent on joint efforts to uphold and improve yields.

Agricultural research depends on inputs from a broad range of disciplines. Isolated researchers are finding it increasingly hard to confront the numerous challenges in increasing food production alone. For instance, agricultural technologies are sensitive to local climate, soil and other attributes. Unlike many medical, mechanical or veterinary innovations that are applicable from one place to another, local adaptation is required for crop technologies. (Reddy,2005).

Electronic access to sources has resulted in productive use of precious time at every stage of research. Some of the significant impacts for scientists include: timely sharing of data and intermediate results for collaboration and discussion by eliciting instant feedback from colleagues and fellow scientists; global participation and narrowing the digital divide; instant interaction with publishers, editors, change agents and potential users; wider scope for publication and dissemination of findings; enhanced, visibility of authors.

The integration of scientific agricultural centers through connection to the Internet has facilitated direct links between scientists and thereby effective collaboration. The exponential growth of electronic sources of information also has greater impact on extension agents and information professionals/librarians.

The Internet has altered the expectations of library users and changed the service perspectives of librarians and other information providers (Lankes, Collins, and Kasowitz, 2000). Increased expectations from the heterogeneous group of patrons involved demand different strategies for searching diverse databases, and this will have the positive impact of increased demand for technical skills to make use of technology.

In some situations the functionally literate farmer is able to interact directly with the scientist. In other cases the information has to be repackaged by the change agent/extension worker to be transferred either directly or thru other channels. More sources will lead to more options for sources, resulting in higher patron expectations, and need for more reliance on new technologies (Tenopir and Ennis, 1998).

The advent of the Internet, reinforced with a variety of web-based information services, has meant that the boundaries between the author, publisher, librarian and user are shrinking. At the same time, the gap between the haves and have-nots is widening. Access to digital sources and, more particularly, collaborative efforts in building up digital sources of scientific information, are therefore a most welcome sign in closing the gap, using the help and expertise of all.

IMPACT OF DIGITAL INFORMATION SOURCES

Digital sources can be used through mediated or unmediated channels. The possible impacts on each of the participants are enumerated.

IMPACT ON THE AGRICULTURAL SCIENTIST

Electronic access to sources will result in productive use of precious time at every stage of research. Some of the significant impacts for scientists include: timely sharing of data and intermediate results for collaboration and discussion by eliciting instant feedback from colleagues and fellow scientists; global participation and narrowing the digital divide; instant interaction with publishers, editors, change agents and potential users; wider scope for publication and dissemination of findings; enhanced visibility of authors.

The integration of scientific agricultural centers through connection to the Internet will facilitate direct links between scientists and thereby effective collaboration.

IMPACT ON PUBLISHER/EDITOR

Publishers can use electronic sources to obtain pertinent information on peer reviewers; to access relevant bibliographic information; to make instant contact with authors and use their web pages; to identify deficient areas of publishing; and to make hyperlinks from references with full texts and multimedia.

The growth of electronic information from publishers offers them an opportunity to reach a worldwide audience in a short time with a variety of information options.

IMPACT ON CHANGE AGENT/INFORMATION PROFESSIONALS

The exponential growth of electronic sources of information is expected to have greater impact on change agents and information professionals.

Although networked electronic sources are accessible to end users, for a variety of reasons such as affordability, availability and acceptability, greater use will be made of the services of change agents and information professionals.

Concurrent with increase in the variety of databases, there will be increased need for onsite as well as remote user instruction, calling for better skills and expertise of the intermediaries.

IMPACT ON USERS

Increased expectations from the heterogeneous group of patrons involved demand different strategies for searching diverse databases, and this will have the positive impact of increased demand for technical skills to make use of technology. As technologies and new information pathwayscontinue to be developed and improved, it will be necessary for users to acquire the skills to access information. Built into these new programs will be a demand by users for ease of access which programmers continually strive to provide. Thus, when farmers in developing countries are first introduced to electronic resources, they will learn quickly that access is as simple as pressing a button on a keyboard or touching a screen with a finger. In addition, for farmers who lack the language skills to comprehend written information, visual and audio enhancements provided on hand held devices will make it easier for extension agents and advisers to present new information to the communities they serve.

CONCLUSION

There are several barriers in communicating information to farmers. Not all the villages in developing countries have internet access. In several countries, the extension agent and farmer ratio is 1:3000.Access to digital sources and, more particularly, collaborative efforts in building up digital sources of scientific information, are therefore a most welcome sign in closing the gap, using the help and expertise of all. Interestingly the mobile phone has given disadvantaged people access to information with which they can build businesses or increase the productivity of their crops in geographically isolated areas and where there are unpaved roads, unreliable public transport, and disruptedlandlines. Several citizen services are now increasingly available on mobile phones, with both the public and private sector playing a role in their delivery.

As the number of people engaged in agriculture around the world decreases, the responsibility for feeding the burgeoning populations, will fall upon the new breed of farmers who have some degree of literacy and education.

For their needs, the four D's – development, documentation, dissemination, and diffusion will be the normative framework for information transfer. Improvements and innovations that are bound to emerge over time, should make agricultural information flow a dynamic and effective method for farmers and urban planners to gain the necessary information that is crucial for their livelihoods and well being.

REFERENCES

American Farm Bureau (2010). Young farmers, ranchers face concerns but express optimism. Retrieved May 20, 2010, from www.fb.org/index.php?fuseaction=newsroom.newsfocusandyear=2010andfile=nr0311c.html

Achleitner, H. K. (1995). Information transfer, technology, and the new information professional.In G.M. Pitkin (Ed.) *The impact of emerging technologies on reference service and bibliographic instruction* (pp.137-149). London: Greenwood Press.

Alene, A.D., and Manyong, V.M.(2007).The effects of education on agricultural productivity under traditional and improved technology in northern Nigeria: An endogenous switching regression analysis. *Empirical Economics,* 32, 141-159.

Alston, J. M., Pardey, P. G., Smith, V. H., and International Food Policy Research Institute. (1999). *Paying for agricultural productivity.* Baltimore: Johns Hopkins University Press.

Alston, J. M., Pardey, G.P.and Roseboom, J. (1998).Financing agricultural research: International investment patterns and policy perspectives. *World Development,* 26, 1057-1071.

American Dietetic Association (1990). Position of the American Dieetetic Association: Domestic hunger and inadequate access to food. *Journal of the American Dietetic Association*, 90, 1437-1441.

Ameur, C. (1994). *Agricultural Extension: A Step beyond the next step.* Washington, D.C: The World Bank.

Anderson, J.R. (1998). Selected policy issues in international agricultural research: On striving for international public goods in an era of donor fatigue. *World Development* ,26, 1149-1162.

Appleton, S. and Balihuta, A. (1996). Education and agricultural productivity: evidence from Uganda. *Journal of International Development*, 8, 415-444.

Arnon, I. (1989). *Agricultural research and technology transfer*. London: Elsevier Applied Science.

Ballantyne, P. (1994).Bridging the gap between practice and policy: Policy management challenges for agricultural information specialists. *Quarterly Bulletin of IAALD,* 39, 24-31.

Barrett, C.B. (2002). Food security and food assistance programs.In*Handbook of Agricultural Economics*, B.L.Gardner and G.C.Rausser (Eds.), vol 2B.*Agricultural and Food Policy*.(Chapter 4, pp.2104-2190). New York: Elsevier Science.

Beintema, N. M., Dias Avila,A. and Pardey,P.G. 2001. Agricultural R&D in Brazil: Policy, investments, and institutional profile. Washington, D.C. and Brasilia: International Food Policy Research Institute and Embrapa

Beintema, N. M. and Pardey, P. G. (2001). Recent developments in the conduct of Latin American agricultural research. Paper prepared for the International Conference on Agricultural Science and Technology, Beijing, November7–9.Retrieved May20,2010from http://www.ifpri.org/sites/default/files/pubs/isnararchive/indicator/pdf/asti_la.pdf

Beintema, N.M.andStads, G.J. (2004). Investing in Sub-Saharan African agricultural research: Recent trends.Vision 2020 Africa Conference Brief 8. Washington, DC: IFPRI,

Beintema, N. M., and Stads, G. J. (2006). Agricultural RandD in sub-Saharan Africa: An era of stagnation. *Background Report for the ASTI Initiative,* Washington, DC: International Food Policy Research Institute.

Benor, D. and Harrison, J. Q. (1977).*Agricultural extension the training and visit system*. Washington,D.C: World Bank.

Berlo, D. K.(1960). *The process of communication; an introduction to theory and practice.* New York: Holt, Rinehart and Winston.

Bernstein, R.H. and Johnson, G.L.(1994). International agricultural research.In. C.J.Arntzen., and E.M.Ritter (Eds.), *Encyclopedia of Agricultural Science (Vol.2*, pp.579-590). San Diego: Academic Press.

Binenbaum, E., Pardey,P.G. and Wright,B.D.(2001). Public private research relationships: The Consultative Group on International Agricultural Research. *American Journal of Agricultural Economics* 83: 748-753.

Blackie,M.J. and Dent,J.B.(1979). *Information systems for agriculture.* London: Applied Science Publishers.

Board on Agriculture and Natural Resources (2002).*Publicly funded agricultural research and the changing structure of U.S. agriculture*: Ad hoc committee report. Washington, D.C: National Academy Press.

Boehlje, M. (1998). Information technology transfer in agriculture :The role of the public and private sectors. In S.A.Wolf(Ed.), *Privatization of information and agricultural industrialization*. New York: CRC Press.

Boone, K., Meisenbach, T., and Tucker, M. (2000).*Agricultural communications: Changes and challenges.*Ames : Iowa State University Press.

Borlaug, N.E. (2001). Feeding the world in the 21st century: The role of agricultural science and technology.AgBioWorld [Online]. Retrieved May10,2010, from http://www.agbioworld.org/biotech-info/topics/borlaug/borlaugspeech.html

Brewer, F. L. (2001). *Agricultural extension systems: An international perspective*. North Chelmsford, MA: Erudition Books.

Britt, D. W. (1997). *A conceptual introduction to modeling: Qualitative and quantitative perspectives*. Mahwah, N.J: Lawrence Erlabaum Associates.

Bruinsma, J. (2003). *World Agriculture: Towards 2015/2030*. London: Earthscan Publications Ltd.

Bunney, M. (1998).Privatization of extension in the United Kingdom: Implications for agriculture and the rural economy. In S.A.Wolf (Ed.).*Privatization of information and agricultural industrialization* (pp.223-238). New York: CRC Press.

Byerlee, D. (1998). The search for a new paradigm for the development of national agricultural research systems. *World Development, 26*, 1049-1055.

CGIAR.(1996). *Agricultural scientists design better future*: new efforts to feed people fight poverty and protect environment in the 21st century. Press release dated October 27, 1996. Retrieved May 20, 2010 from http://www.worldbank.org/html/cgiar/press/25rel.html

CGIAR. (2003). The CGIAR at 31: Celebrating its achievements, facing its challenges. *Precis 232*. Retrieved May 20, 2010 from http://lnweb90.worldbank.org/oed/oeddoclib.nsf/DocUNIDViewForJavaSearch/D01D16C0D67B0F3085256D56005AE040/$file/Precis_232_CGIAR.pdf

Chema, S., Gilbert, E. and Roseboom,J. (2003). A critical review of key issues and recent experiences in reforming agricultural research in Africa.*ISNAR research report no. 24.* The Hague, The Netherlands: .International Service for National Agricultural Research.

Christison, B. (2000). Impact of globalization on family farm agriculture.*Motion Magazine*, August 1. Retrieved May, 20, 2010 from http://www.inmotionmagazine.com/bcbrasil.html

Core, P. (2009).*A retrospective on* rural RandD *in* Australia. *Retrieved May 20,2010, from* http://www.daff.gov.au/__data/assets/pdf_file/0020/1435 331/rd-retrospective-report.pdf

Crook, F. W. (1992). China's extension system. *Journal of Extension, 30*, 30-31.

De Hann., C., Steinfeld, H .and Blackburn, H.(1998). *Livestock and the environment; finding a balance.*Commission of the European Communities and the World Bank.

Delman, J. (1988). *The agricultural extension system in China*. London: ODI, Agricultural Administration Unit.

Doctors, S. I. (1969).*The role of federal agencies in technology transfer*. Cambridge, MA: M.I.T. Press.

Echeverría, R. G. (1998). Agricultural research policy issues in Latin America: An overview. *World Development, 26*, 1103-1111.

Etzkowith, H. (2001). Beyond the Endless Frontier.In Wolf, S. and Zilberman, D. (Eds).*Knowledge Generation and Technological Change :Intitutional Innovation in Agriculture* (pp.3-26). Boston: Kluwer Academic.

Evanson, R.E.and Douglas G.B. (2002).Contributions of national agricultural research systems to crop productivity. In Gardner, B.L. and Rausser, G.C.(Eds), *Handbook of agricultural economics, vol. 3: Agricultural development: Farmers, farm production and farm markets*, (Chap. 46. pp.2419-2459). New York : Elsevier Science.

Evenson, R.E. and Gollin, D. (2003).Assessing the impact of the green revolution, 1960 to 2000.*Science,* 300 (5620), 758-762.

Evenson, R.E. and Gollin, D. (2003).*Crop Variety improvement and its effect on productivity: The impact of international Agricultural research.* Oxford: CAB Publishing.

Evenson, R.E. (2001).Economic impacts of agricultural research and extension. In Gardner, B.L. and Rausser, G.C.(Ed.). *Handbook of Agricultural Economics*(vol. 1A. Chap.11. pp.574-616).New York: Elsevier Science.

FAO. (2003). Ministerial meeting on "water for food and agriculture": Issue paper prepared by Ministry of Agriculture, Forestry and Fisheries of Japan and Food and Agriculture Organization of the United Nations Kyoto, Japan. Retrieved May 20, 2010, from ftp://ftp.fao.org/agl/aglw/ docs/kyotoissuepaper.pdf

FAO. (2005). Committee on World Food Security, *Special Eventon green revolution in Africa: Background document* prepared by SDR. Retrieved May 20, 2010, from ftp://ftp.fao.org/docrep/fao/meeting/009/j5421e.pdf

FAO/UNDP. (1980). *China : The agricultural trainingsystem : Report on an study tour to the people's republic of china 5 October to 2 November 1978*. Rome: Food and Agriculture Organization of the United Nations.

FAO/UNESCO/ILO.(1970). *World conference on agricultural education and training report*.Vol 1.Proceedings of the conference. Copenhagen, Denmark. Rome: FAO.

Farquhar, R.N. (1972). *Australian Agricultural Extension Conference, 1962, Review, Papers and Reports* (pp.114-117).

Frison, E. (1999). The establishment of Global Programmes for Commodity Chains Discussion Paper1 for the GFAR Steering Committee meeting of 23 May 1999.Retrieved May 20, 2010, from http://www.fao.org/docs/eims/upload/216239/Establishment_GPP_for_CC.pdf

Gamble, W.K. (1997). The Globalization of science: Agricultural research in developing countries. In C. Bonte-Friedheim and K.Sheridan (Ed.).*The Globalization of Science: The Place of Agricultural Research* (pp.41-45). The Hague: ISNAR.

Graff, G. (2002). *Generating and trading biological innovations in agriculture*. Ph.D. Dissertation.Berkeley,CA: University of California.

Hardin, L.S. (1994). Whence international agricultural research? *Food Policy* 19: 561-567.

Harrison, P., Pearce, F., and American Association for the Advancement of Science.(2001). *Population and biodiversity*. Berkeley, CA: University of California Press. Retrieved May15, 2010, from http://atlas.aaas.org/pdf/159-62.pdf.

Hayami, Y. and Ruttan,V.W. (1988). Induced technical change in agriculture, In S. M. Capalbo and J.M. Antle (Eds.), *Agricultural productivity measurement and explanation* (pp.247-288). Washington, DC: Resources for the Future.

Henriquez, P. and Franca, Z.(2004). *Innovate or die: Learning to survive in the new agricultural research environment*.Briefing Paper 77. The Hague: ISNAR.

Hu, R., Yang, Z., Kelly, P., and Huang, J. (2009). Agricultural extension system reform and agent time allocation in china. *China Economic Review (1043951X), 20*(2), 303-315. doi:10.1016/j.chieco.2008.10.009.

Huffman, W. E. and Orazem,P.F.(2001).Agriculture and human capital in economic growth. In B.L. Gardner and G.C. Rausser (Eds.).*Handbook of*

Agricultural Economics, vol. 3: *Agricultural Development: Farmers, farm production and farm markets* (chap. 43, pp.2281-2341). New York: Elsevier Science.

Huang, J., R. Hu and Rozelle,S. (2003). *Agricultural research investment in China: Challenges and prospects.* Beijing: Financial and Economic Press.

IAASTD.(2008). *International Assessment of Agricultural Knowledge, Science and Technology for Development Synthesis Report.* Retrieved May 15, 2010, from http://www.agassessment.org/index.cfm?Page= doc_libraryandItemID=14I

ICARDA. (1999). *The national agricultural research systems in west Asia and North Africa region.* Aleppo, Syria: ICARDA.

Jackson, S. W., Hopper, G. M., and Clatterbuck, W. K. (2004).Developing a national web-based learning center for natural resource education.*Journal of Extension, 42*(1) (online).

Janssen, W. and Kassam,A. (2004). A regional approach to setting research priorities and implementation: towards satisfying national, regional and global concerns. *Journal of Agricultural and Food Information,*5(2), 67-100.

Jain, H.K. (1999). Time for resurgence of public sector agricultural research.*Biotechnology and Development Monitor*, 39,24.

Janssen, W. and T. Braunschweig, T.(2003). Trends in the organization and financing of agricultural research in developed countries: Implications for developing countries. *Research Report 22.* The Hague: ISNAR.

Katz, S., and Ben-David, J. (1975).Scientific research and agricultural innovation in Israel. *Minerva, 13*, 152-182.

Kerka, S. (1998).Extension today and tomorrow.*Trends and Issues Alerts.*ERIC Clearinghouse on Adult, Career, and Vocational Education.(ERIC Document Reproduction Service No. ED425335).

Knight, J. , Weir, S., and Woldehanna, T.(2003)The role of education in facilitating risk-taking and innovation in agriculture. *Journal of Development Studies,* 39, 1-22.

Kerr, J. and Kolavalli,,S. (1999). Impact of agricultural research on poverty alleviation: conceptual framework with illustrations from the literature. EPTD discussion paper no. 56. Environment and Production Technology Division and Food Consumption and Nutrition Division. Washington DC: International Food Policy Research Institute.Rereieved May 24, 2010, from http://www.ifpri.org/divs/eptd/dp/papers/eptdp56.pdf.

Krishnakumar, Dhanya. (2007).Mobilising growthin rural India. *Times of India / Times; Business,* 3 Sept. 2007, p. 24..

Landau, H. B., Maddock, J.T., Shoemaker,F.F., and Costello,J.G. (1982). An information transfer model to define information users and outputs with specific application to environmental technology.*Journal of the American Society for Information Science,* 33, 82-91.

Lankes, R., David, J.W., Collins III., and Kasowitz,A.S (Eds). (2000). *Digital reference service in the new millennium*: Planning, management, and evaluation. New York: Neal-Schuman.

Latour, M. A., and Meunier, R. A. (1999). Transferring poultry information to the public using the internet: AvianNet@ purdue university.*Journal of Extension, 37*(5) (Online).

Lele, U., and Coffman, R. (1995).*Global Research on the Environmental and Agricultural Nexus for the 21st Century:* A Proposal for Collaborative Research among US Universities, CGIAR Centres, and Developing Country Institutions. Report of the Taskforce on Research Innovations for Productivity and Sustainability. Gainesville, FL: Office of International Studies and Programs, University of Florida.

Loebenstein, G. and Putievsky, E. (2007).Agricultural Research in Israel.In G.Loebenstein.andG.Thottappilly (Eds.) *Agricultural Research Management.* (pp.357-365). Dordrecht, The Netherlands: Springer.

Lockeretz, W., and Anderson, M. D. (1993).*Agricultural research alternatives*. Lincoln: University of Nebraska Press.

Maredia, M. K., and Raitzer, D. A. (2006). CGIAR and NARS partner research in sub-Saharan Africa: Evidence of impact to date.Washington D.C: CGIAR.

McNamera, Robert S. 1977. *Address to the Massachusetts Institute of Technology.* Cambridge. Washington : International Bank for Reconstruction and Development.

McCue, J., Craycraft, C., Dunham, T., Fretz, T., McGeachin, R., Wilson, P., and Young, E. (2005), Leadership Council for Agricultural Information and Outreach. Subcommittee on Content"*Stocking the Shelves*" Report. Lexington, KY. Retrieved May 10,2010, from http://usain.org/ WhitePaperFinal.pdf

Michelsen, H., and Hartwich, F. (2004) *University-based Agricultural Research: A Comparative Study in Sub-Saharan Africa* — Research Report 26. The Hague: International Service for National Agricultural Research.

Miller, R.J., Hanson, J., Fretz, T., and Weismiller, R.D.(2004). Science and education: Necessary conditions for successful agriculture and rural development*Outlook on Agriculture*, 33, 55-58.

Mruthyunjaya and Ranjitha, P. (1998). Indian agricultural research system: Structure, current policy issues, and future orientation. *World Development,* 26, 1089-1101.

Mudahar, M.S., Jolly, R.W., and Srivastava, J.P. (1998).*Transferring agricultural research systems in transition economies: The case of Russia.* World Bank Discussion Paper No. 396. Washington D.C: The World Bank.

National Science Foundation , Division of Information Science and Technology (1977). *A Chart book of indicators of scientific and technical communication in the United States.*Washington, D.C.: NSF.

National Agricultural Statistics Service.(2009). *Farm computer and ownership* August 2009.Retrieved May 20, 2010, from http://usda.mannlib.cornell.edu/usda/current/FarmComp/FarmComp-08-14-2009.pdf

Neelameghan, A. and Chester, G. (2010). Use of mobile and wireless communication:With special reference to rural India.*Internet-ZeitschriftfürKulturwissenschaften*(Internet Journal of Indegenous Studies) 17 Section 8.2.

Pardey, P.G., Roseboom, J., and Beintema, N.M.. 1995. *African agricultural research:Three decades of development. I*SNAR Briefing Paper No. 19, The Hague: ISNAR..

Pardey, P. G., Roseboom,J., and Beintema,N.M. (1997). Investments in African agricultural research.*World Development*, 25, 409-423.

Paarlberg, R. (2000). The Global Food Fight.*Foreign Affairs* 79,24-38.

Pates, M. (2004).*Extension service still vital: Remains agent for change. Retrieved March 10, 2010, from* http://www. in-forum.com/specials/century/jan3/week13.htm

Pingali, P. L., and. Traxler,G.(2002). Changing locus of agricultural research: Will the poor benefit from biotechnology and privatization trends? *Food Policy,* 27, 223-238.

Plucknett, D.L., and Smith,N.J.H. 1984. Networking in international agricultural research.*Science* 225(4666), 989-993.

Pollack, A. (2001). The Green revolution yields to the bottom line. *New York Times,* May 15.Retrieved May 20, 2010, from http://www.nytimes.com/2001/05/15/science/the-green-revolution-yields-to-the-bottom-line.html

Qijie, G., and Chuanhong, Z. (2008). Agricultural technology extension system in china: Current situation and reform direction. *Management Science and Engineering, 2*, 47-58.

Radcliffe, J. C., and Clarke, A. E. (1996).Rural research in Australia.*Science, 273* (5275), 612.

Randhawa, M.S. (1979).*A history of the ICAR*. New Delhi: Indian Council of Agricultural Research.

Ramos,M.M., Alvare,L.M., Shelton,P. &Besemer,H.(2009).The CGIAR Virtual Library: A Vital Tool to Enhance Access to Agricultural Research.*Journal of Agricultural & Food Information*, 10, 27-36.

Ranganathan, S.R. (1931).*The five laws of library science*.Madras: Madras Library Association.

Reddy, D. E. (2004). Globalization of agricultural science and education: An overview of texasAandM university international agricultural programs. *Quarterly Bulletin of the International Association of Agricultural Information Specialists, 49*, 24-29.

Reddy, D.E. (2005). Changing trends in international agricultural research: Implications on information transfer. *Quarterly Bulletin of the International Association of Agricultural Information Specialists*, 50: 1/2 (2005): 24-29.

Rivera, W.N. (2006). Agricultural knowledge and development in a new age and different world.*Journal of International Agricultural and Extension Education*, 13(2) 57- 67.

Rogers, E.M. (1962). *Diffusion of innovations*. New York: Free Press of Glencoe.

Rogers, E. M. (1983). *Diffusion of innovations*. New York: Free Press Macmillan.

Röling, N. G. (1988). *Extension Science: Information Systems in Agricultural Development*. New York ,NY: Cambridge University Press.

Roseboom, J., Pardey, P. G., and Beintema, N. M. (1998).*The Changing Organizational Basis of African Agricultural Research*, EPTD Discussion Paper No 37. Washington, DC: IFPRI.

Runge, C. F., Benjamin, S., Pardey, P.G. and Rosegrant,M.W. (2003). *Ending hunger in our lifetime: Food security and globalization*. Baltimore: Johns Hopkins University Press.

Russell, E. J. (1966). *A history of agricultural science in Great Britain, 1620-1954*.London: George Alien and Unwin.

Ruttan, V.W. (1989). International agricultural system.In J.L.Compton (Ed.) *The transformation of international agricultural research and development (pp.*173-205). Boulder: LyneeRienner Publishers.

Schmitt, M.A., Durgan, B.R. and Iverson, S.M. (2000).Impact assessment and participant profiles of extension's education programs for agricultural chemical/seed retailers and crop advisors.*Journal of Extension*, 38, 8.

Shand, H. (2001). *Private sector agricultural research*: Research priorities of agribusiness and how they are changing farmers' lives. ETC Group. http://www.evb.ch/cm_data/Hope%20Shand.pdf

Sharma, S. D. (2003). Agricultural information management and systems and services. *Quarterly Bulletin of the International Association of Agricultural Information Specialists, 48*(1/2), 36-43.

Shashok. K. (2001).Author's editors: Facilitators of science information transfer. *Learned Publishing,* 14: 113-121.

Singh, K.(2000). Education, technology adoption and agricultural productivity.*Indian Journal of Agricultural Economics*, 55, 473-489.

Slater, S.W. (1951). The organization of agricultural research in Great Britain.*Journal of the Science of Food and Agriculture*,2, 337-41.

Swaminathan, M.S.,Sneep, J., and. Hendriksen,T.A.J.(1979). Perspspectives. In J. Sneep , A. J. T. Hendriksen and O. Holbek (Eds) *Plant breeding perspectives* (pp.396-429). Wagenigen: Center for Agriculture and Documentation.

Swaminathan,M.S. and Ramanujam,S. (1980). *Science and agriculture : M.S. Swaminathan and the movement for self-reliance.* Jodhpur. Arid Zone Research Association of India.

Swanson, B. (2006). The changing role of agricultural extension in a global economy. *Journal of International Agricultural and Extension Education, 11*, 5-17.

Stads, G. J., and Beintema, N. M. (2009). Public agricultural research in Latin America and the Caribbean:Investment and capacity trends, ASTI Synthesis Report. Washington D.C: International Food Policy Research Institute.

Tacchini, J. (1995). Education of the agricultural population to improve productivity and quality of life.*Revista de la Facultad de CienciasAgrarias, Universidad Nacional de Cuyo,* 27(1), 72.

Tenopir, C. and Ennis, L. (1998).The digital reference world of academic libraries.*Online* 22(6): 84-6.

Thirtle,C., Lin,L., and Piesse,P.(2003). The impact of research-led Agricultural Productivity Growth on Poverty Reduction in Africa, Asia and Latin America.*World Development*, 31, 1959-1975.

Tribe, D. (1994).*Feeding and greening the world.* Oxford: CAB International.

Trigo, E. J. (1987).Agricultural research organization in the developing world : Diversity and evolution. In V. W.Ruttan and C.E.Pray (Eds) *Poloicy for Agricultural research* (pp.251-281). Boulder : Westview Press.

UN Millennium Project. (2005). *Halving hunger: it can be done*. New York: The Earth Institute at Columbia University. Retrieved May 20, 2010, from http://www.unmillenniumproject.org/documents/HTF-SumVers_FINAL.pdf

United Nations (UN). 2000. *UN Millennium Development Goals*.Rerieved May24, 2010, from http://www.un.org/millenniumgoals/

Weintraub, Irwin. (2000). The impact of alternative presses on scientific communication. *International Journal on Grey Literature*, 1, 54 – 59.

Winrock International. (1978). *The role of ruminants in support of man*. Little Rock, Arkansas:Winrock International.

Wolf, S.A. (1998). *Privatization of information and agricultural industrialization*. New York: CRC Press.

Wolf, S. A., and Zilberman, D.(2001). *Knowledge generation and technical change*: Institutional innovation in agriculture. Boston: Kluwer Academic Publishers.

World Bank.(2008).*The food crisis in Latin America and the Caribbean*.Retrieved May 10,2010 from http://go.worldbank.org/X6NTCGFEQ0

Wortman, S. and Cummings, R. W. (1978).*To feed this world: The challenge and the strategy*. Baltimore: Hopkins University Press.

INDEX

M

N

O

P